Clinical Aspects of Calcium Entry Blockers

Progress in Basic and Clinical Pharmacology

Vol. 2

Series Editors
P. Lomax, Los Angeles, Calif.
E. S. Vesell, Hershey, Pa.

KARGER

Basel · München · Paris · London · New York · New Delhi · Singapore · Tokyo · Sydney

Clinical Aspects of Calcium Entry Blockers

Volume Editor
P. A. van Zwieten, Amsterdam

14 figures, 1 color plate and 15 tables, 1989

KARGER

Basel · München · Paris · London · New York · New Delhi · Singapore · Tokyo · Sydney

Progress in Basic and Clinical Pharmacology

Library of Congress Cataloging-in-Publication Data
Clinical aspects of calcium entry blockers
(Progress in basic and clinical pharmacology; vol. 2)
Includes bibliographies and index.
1. Calcium – Antagonists – Therapeutic use – Testing. 2. Blood-vessels – Effect of Drug on. I. Zwieten, P.A. van. II. Series.
[DNLM: 1. Calcium Channel Blockers – pharmacology. 2. Calcium Channel Blockers – therapeutic use. WI PR666GK v. 2 / QV 276 C6406]
RM666.C243C55 1988 616.1'061 88-29625
ISBN 3-8055-4858-3

Drug Dosage

The authors and the publisher have exerted every effort to ensure that drug selection and dosage set forth in this text are in accord with current recommendations and practice at the time of publication. However, in view of ongoing research, changes in government regulations, and the constant flow of information relating to drug therapy and drug reactions, the reader is urged to check the package insert for each drug for any change in indications and dosage and for added warnings and precautions. This is particularly important when the recommended agent is a new and/or infrequently employed drug.

Printed by Konkordia Druck GmbH, Bühl/Baden (FRG)
ISBN 3-8055-4858-3

Contents

Treatment of Myocardial Ischemia with Calcium Entry Blockers

Treatment of Hypertension with Calcium Antagonists

Clinical Aspects of Calcium Entry Blockers in Cardiomyopathy, Primary Pulmonary Hypertension and Raynaud's Phenomenon

Treatment of Migraine with Calcium Entry Blockers

Treatment of Neurological Disorders Except Migraine with Calcium Antagonists

Preface

Calcium antagonists (CAs) have drawn the attention as drugs that are both therapeutically relevant and of great fundamental importance. Their discovery has greatly deepened our insight into the fundamental processes of cellular calcium metabolism, and the concept 'specific calcium channels' in the cell membrane would probably not have been submitted without the availability of the CAs as selective blockers of these channels. Most review papers and monographs on CAs have emphasized either their fundamental mechanisms or, clinically, the treatment of one particular disease with these therapeutic agents.

Reviews or books covering the broader spectrum of the therapeutic applicability of the CAs are rare. It thus seemed of interest to ask experts in their various fields to present the state of the art of the CA drugs as therapeutic means in the most important clinical conditions where they are applied at present. After two introductory chapters of a more basic character, the role of CAs in various cardiovascular and neurological disorders is described in some detail, with emphasis on the clinical aspects but without neglecting the basic mechanisms involved, as far as relevant.

We have limited the discussion in this book to those diseases where the therapeutic role of CAs is substantiated by clinical trials, however modest they may sometimes be. Those diseases where only case reports and clinical impressions are available with respect to the possible therapeutic activity of CAs are not included here, but the reader may consult the recently published volume (No. 552) of the *Annals of the New York Academy of Sciences* on Calcium Antagonism edited by P.M. Vanhoutte, L. Paoletti, and S. Govoni.

I was fortunate in finding such highly competent authors for the various chapters and subjects and I wish them to be assured of my gratitude for their contributions.

P.A. van Zwieten

Van Zwieten PA (ed): Clinical Aspects of Calcium Entry Blockers.
Prog Basic Clin Pharmacol. Basel, Karger, 1989, vol 2, pp 1–20

Pharmacological Background of the Therapeutic Activities and Side Effects of Calcium Antagonists

P. A. van Zwieten

Divisions of Pharmacotherapy and Cardiology, Academic Medical Centre, University of Amsterdam, The Netherlands

Introduction

Calcium antagonists (CAs) or calcium entry blockers are considered as a breakthrough in the drug treatment of cardiovascular disease, while their importance as tools in fundamental research is generally recognized as well. More recently, certain types of CAs have been studied as potentially useful drugs in the treatment of neurological disorders like, for instance, migraine, cerebral ischemia and hypoxia, certain forms of vertigo, epilepsy and subarachnoidal hemorrhage. Simultaneously, the number of drugs has grown steadily. It seems correct to consider the introduction of CAs as one of the major innovations both in pharmacotherapy and in experimental pharmacology.

It is the aim of the present chapter to review the basic properties and potential application of these drugs in various disorders, with emphasis on the pharmacological background of the therapeutic activity and adverse effects. Organ specificity, an obvious characteristic of the CAs will be discussed as well.

Nomenclature and Definitions

According to Fleckenstein's [1] classical and widely accepted definition, a CA is a drug which inhibits the slow inward calcium current (fig. 1) in a dose-dependent manner, whereas a second requirement is that this effect can be reversed by calcium ions. Accordingly, the drug-induced inhibition

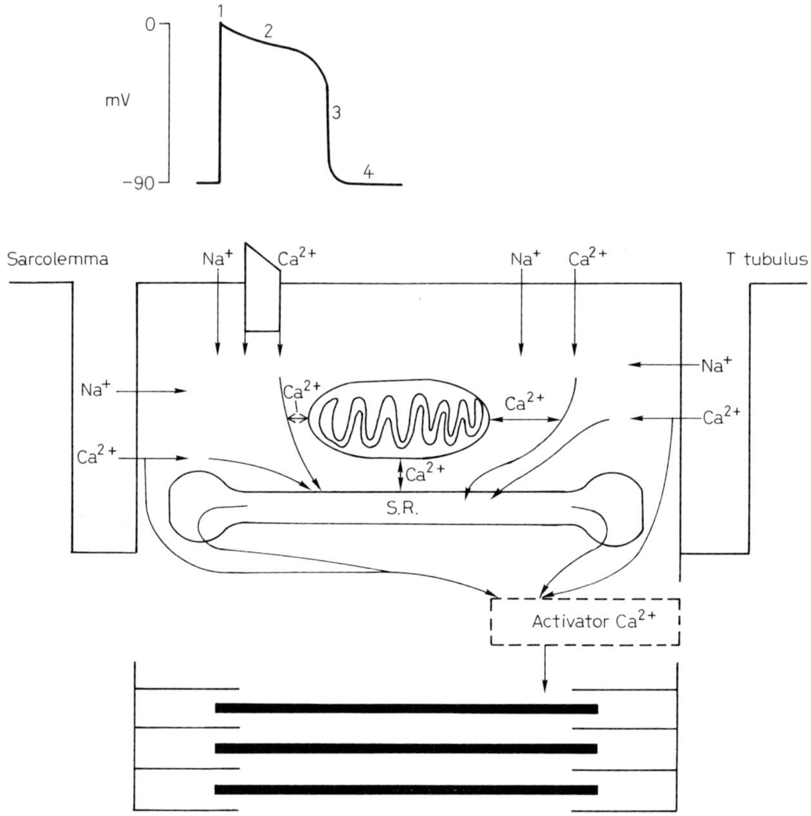

Fig. 1. Effect of CAs on a cardiac cell. Top: typical cardiac action potential. The calcium (slow) inward current flows during the characteristic plateau phase (phase 2) of the action potential. This calcium influx is selectively inhibited by CAs. Activation of the sarcoplasmic reticulum (SR) and other cellular calcium pools occurs via Ca^{2+} and Na^{+} ions which flow into the cell. The SR and other pools donate activator Ca^{2+} ions which stimulate the contractile proteins. The presence of tubular systems (invaginations), which are characteristic of cardiac tissues, result in considerable enlargement of the cellular surface, thus enabling an effective influx of Na^{+} and Ca^{2+} ions. Inhibition of the calcium inward flux by a CA causes diminished activation of the contractile proteins.

of the slow inward current, which is known to be carried by calcium ions, obeys the laws of competitive inhibition. In this connection, the potency of the CAs draws the attention: a single molecule of such a drug is able to prevent the influx of several thousands of calcium ions. From a pharmacological point of view the term "calcium antagonism" may be subject to fundamental criticism. In the pharmacological sense of the word, an antagonist is a drug which inhibits the effect of another drug like, for instance, atropine versus acetylcholine, or propranolol versus isoprenaline. Verapamil and other "CAs", however, do not really inhibit the *effects* of calcium ions; rather they inhibit the passage of calcium ions through specific channels in cell membranes. For this reason the term "calcium entry blocker", proposed by Vanhoutte and Bohr [2] would seem more appropriate. An equivalent, alternative expression would be "slow-channel inhibitors" or "slow-channel blockers" or "calcium channel blockers", as proposed by Nayler and Poole-Wilson [3] and Nayler [4]. The designation 'calcium entry blockers' is also frequently encountered, and would be preferable for a variety of reasons. Since, however, the term *calcium antagonists* (CAs) has been maintained in the recently issued WHO classification [5], we shall henceforth use it throughout.

The term 'calcium overload blockade' has recently been suggested by Borgers [6] and ourselves [7, 8] for the drug-induced inhibition of the overloading of the cell under pathological, ischemic conditions. This term, implicating the nomenclature 'calcium overload blockers' for certain calcium entry blockers, like, for instance, flunarizine, is still subject to debate and will not be generally employed in the present survey.

It should be realized that several other compounds besides the generally recognized CAs can block to some degree the specific calcium channel along with a variety of other pharmacological activities. For instance, various local anesthetics and other drugs with local anesthetic activity, like lidocaine, quinidine or propranolol, are known to impair the slow-calcium channel besides their effective blockade of the fast-sodium channel, the latter being the basis of their local anesthetic potency. Nevertheless, it would be quite inappropriate to call lidocaine, quinidine, propranolol or procainamide CA or calcium entry blockers. The classification should preferably be governed by the predominant action of the drug: accordingly, the aforementioned local anesthetic agents should be classified as sodium channel blockers and *not* as CAs, in spite of their demonstrable calcium antagonistic activity.

Chemistry

As is obvious from figure 2, the CAs are a most heterogeneous series of compounds. Accordingly, several different subgroups should be distinguished.

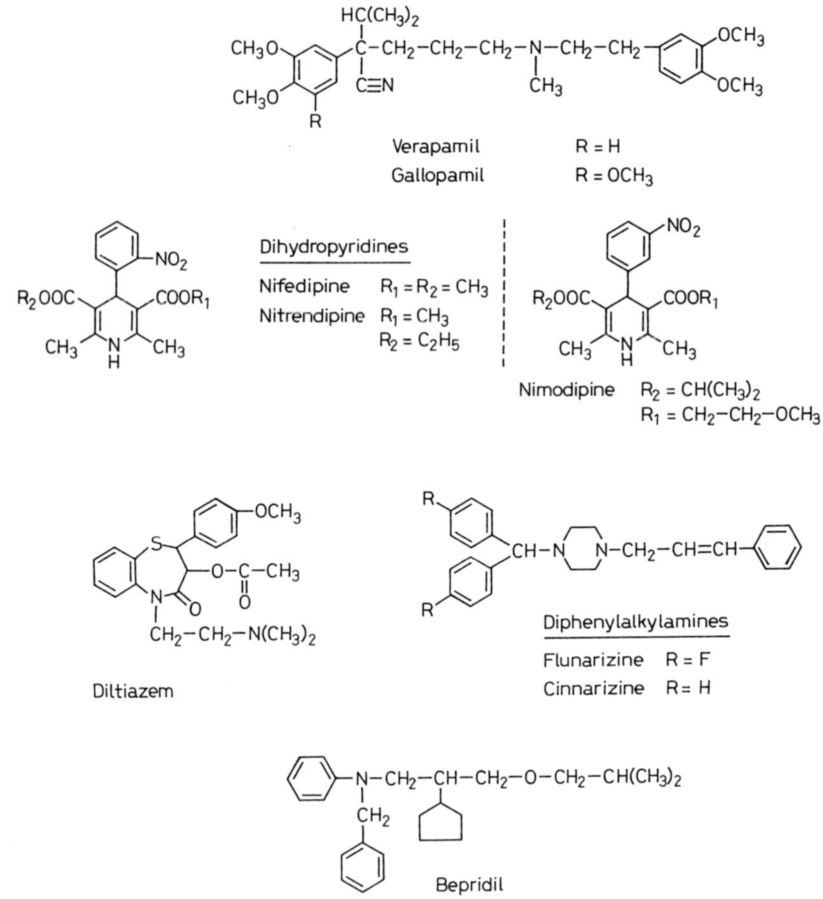

Fig. 2. Structural formulae of CAs. Note heterogeneity of structures. A relationship between chemical structure and pharmacological activity has so far been discovered only within the 1,4-dihydropyridines (nifedipine and related compounds) and within the diphenylalkylamine series (flunarizine and related drugs). Note the deviating structures of diltiazem and bepridil. Flunarizine and nimodipine are the CAs used most frequently in neurological disorders.

The oldest compounds, i.e., verapamil and gallopamil (D 600) are at least chemically related to the β-blockers, although these compounds do not possess significant β-adrenoceptor-antagonistic activity. Shortly after their discovery, they were even considered to be β-blockers, but this view soon proved erroneous.

Chemically speaking, diltiazem, as a benzothiazepine derivative, shows some resemblance to the benzodiazepines, although there is no pharmacological similarity. The dihydropyridines form a rather extensive group of CAs for which a quantitative structure-activity relationship has been proposed [9–11]. Their difference from verapamil and diltiazem is obvious.

Flunarizine and cinnarizine are diphenylalkylamines (fig. 2). They all possess a voluminous lipophilic group (the diphenyl moiety) separated by a 2- or 3-carbon atom chain or by a piperazine ring from an alkaline-substituted nitrogen atom. It seems of interest to note that the structure-activity requirements within the diphenylalkylamines follow those of calmodulin inhibitors [12].

All calcium channel blockers so far developed are lipophilic compounds where the following order of potency holds: flunarizine, cinnarizine >> verapamil, diltiazem > dihydropyridines (nifedipine, etc.).

As a consequence of their lipophilicity, most of the CAs developed so far can penetrate into the brain by crossing the blood-brain barrier. Those used in neurology, in particular flunarizine and nimodipine, are known to achieve significant concentrations in brain tissue and in the cerebrospinal fluid. Owing to their lipophilic character, most CAs also achieve high concentrations in the cell membranes of various central and peripheral tissues.

Radioligand-Binding Studies

The availability of radioactively labeled CAs with a high specific activity, in particular [^{3}H]-nitrendipine, has opened the possibility to perform radioligand-binding studies and hence characterize the specific calcium ion channels in more detail.

Radioligand-binding studies have been carried out mainly with radioactively labeled dihydropyridine derivatives, such as [^{3}H]-nitrendipine, in rat brain membranes [13], and in a later stage also with cardiac and skeletal muscle preparations. For the dihydropyridines, three different binding sites have been identified which might lead to the subdivision of the calcium channel blockers into three subgroups with different affinities for these binding sites [13]. However, this subdivision may be different from that based upon pharmacological techniques and it remains an open question whether the binding sites found in radioligand binding studies are identical to those associated with calcium channels. Nevertheless, it is of interest to note that the dihydropyridine calcium channel blockers can effectively displace the dihydropyridine radioligands from their specific binding sites in a specific manner, suggesting that these binding sites and calcium channels may somehow be associated [14, 15].

It remains difficult to judge the functional relevance of radioligand-binding data obtained in skeletal muscle homogenates, a preparation frequently employed to study the binding affinity of CAs. Although a substantial affinity of CAs for binding sites in skeletal muscle tissue is readily demonstrable, it remains difficult to accept that functionally relevant calcium channels are involved, since CAs do not influence the function of skeletal muscle preparations whatsoever. Nevertheless, radioligand-binding data obtained in skeletal muscle preparations are frequently assumed to be relevant to the interaction between CAs and calcium channels.

Calcium Channels

Radioligand-binding studies have led to a certain view on the nature of the calcium ion channel, which may be summarized as follows: the calcium channel is assumed to span the entire membrane owing to its dimensions and oligomeric structure. It is probably made up of subunits, which are interconnected by disulfide bonds. The subunits also possess free SH groups, which appear to be essential for the function as an ion channel. The detailed structure of the ion channel remains to be clarified, but so far a channel protein with a molecular weight of 180,000 has been identified. The channel proteins are glycoproteins, and the drug receptor sites are depicted as if on different elements or on the same element. Control sites, where phosphorylation is probably an important process, are assumed to be governed by hormones and neurotransmitters. It should be noted that chemically very different CAs all show obvious affinity for the calcium channel. This apparent discrepancy may be explained by the assumption, that different binding sites exist per individual ion channel, which are mutually connected and influence each other by allosteric interactions [for reviews on this subject, see ref. 13–15]. With respect to the influence of CAs on the ion channels it is assumed, that a drug/receptor (binding site) interaction occurs at a contact domain of distinct elements which forms and decays, for instance as a function of state, that is opened and closed.

Over the years, the concept of specific calcium channels has become more refined and it is now generally assumed that these channels are by no means homogeneous, at least three different subtypes being distinguished at present. Originally, a clear distinction was made between potential (or voltage)-operated (POCs) and receptor-operated channels (ROCs).

The major effect of the CAs was assumed to occur at the level of the POCs, whereas the influence of CAs on ROCs is still subject to some debate. More recently, the nomenclature L-, T- and N-channels has been introduced [16] for the distinction of the three different calcium channels, which are demonstrable in various tissues. This field is advancing at a rapid pace and is subject to conciderable change in short time intervals.

Pharmacology

Before discussing pharmacological differences between the various CAs, it seems of interest to summarize briefly the current experimental methods employed to demonstrate their specific pharmacological activity.

CAs should preferably be assessed in isolated preparations where calcium fluxes are measured as such, for instance by means of ^{45}Ca [17]. Useful results have been obtained with isolated vascular smooth muscle preparations, although the technique is too complicated for screening purposes. In heart muscle preparations, unsatisfactory results are obtained [17]. For these reasons, the following two methods are employed and generally recognized in spite of their drawbacks:

(1) Electrophysiological methods, using the voltage clamp technique, mainly in isolated cardiac tissue preparations, allow one to visualize the transmembranous calcium influx through the slow channels, which is impaired in a specific manner by calcium channel blockers [1, 18, 19].

(2) In isolated blood vessel preparations depolarization induced by excess potassium causes a transmembranous influx of calcium ions and hence vasoconstriction. These processes are inhibited specifically by CAs [18–20].

For a detailed discussion of various other models used to assess calcium entry blockade, see Bou et al. [20].

The blockade of the aforementioned specific calcium channels remains the one and common mechanism of all CAs, notwithstanding their obviously different chemical and other characteristics. Events at the cellular membrane and their impairment by CAs have been extensively reviewed [16, 21, 22]. Our group [23–25] has shown that the vasoconstriction induced

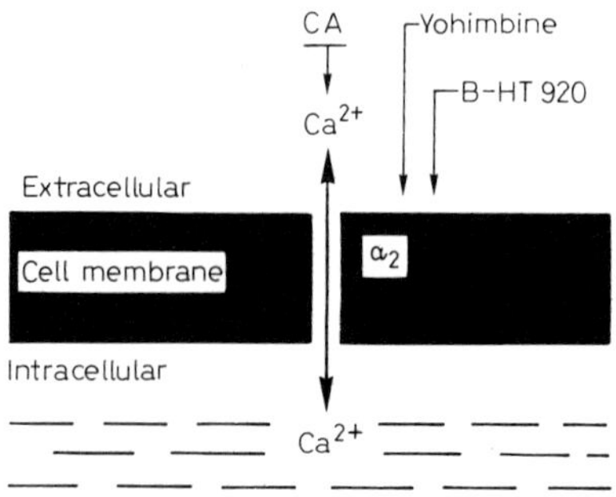

Fig. 3. Schematic representation of the interaction between α_2-adrenoceptor-stimulated vasoconstriction and CAs. Stimulation of the postsynaptic α_2-adrenoceptor in vascular smooth muscle with an α_2-agonist, like for instance the selective α_2-stimulant B-HT 920, is associated with a transmenbranous influx of calcium ions, which contributes to the excitation of the contractile proteins. CAs inhibit this influx of calcium ions and hence impair vasoconstriction. However, CAs do not directly interfere with the α_2-receptor or with the formation of the agonist-receptor complex. Data from the author's department [23–26].

by the stimulation of postsynaptic α_2-adrenoceptors in resistance vessels by appropriate agonists is selectively impaired by calcium channel blockers. This effect is strictly correlated to calcium channel blockade, reflecting the initiation of a calcium inward flux sensitive to CAs on the formation of the α_2-adrenoceptor-agonist complex (fig. 3). This finding suggests that the receptor involved in triggering ROC is probably an α_2-adrenoceptor. Evidence is accumulating that α_1-adrenoceptors may also be involved in certain tissues and species, although probably less consistently than the α_2-adrenoceptor. The role of α-adrenoceptors in triggering the opening of ROCs is probably of major fundamental importance. It may also contribute to the vasodilator activity which is characteristic of virtually all CAs. Accordingly, the calcium influx triggered by α-adrenoceptor stimulation is inhibited by the CAs, hence reducing the activation of contractile proteins. It should be recalled, that both α_1- and α_2-adrenoceptors are stimulated by endogenous catecholamines; the attenuation of the sequelae of this stimulatory activity by CAs leads to vasodilatation. Accordingly, it seems justified to assume that there exists a refined interaction between CAs and the peripheral sympathetic nervous system [for reviews, see ref. 23–25]. Very recently, we have presented experimental findings favoring the view that besides α-adrenoceptors, the angiotensin II receptor also plays a mediator role in triggering the opening of ROCs [26]. Accordingly, angiotensin II receptors may also be involved, though indirectly, in the vasodilator potency of the CAs. Furthermore, this interaction between CAs and angiotensin II receptors should be considered when CAs and converting enzyme inhibitors are administered simultaneously in antihypertensive treatment.

Organ Specifity

The heterogeneity of the various CAs is also reflected by considerable differences in organ specifity. This is a very important issue which may lead to potentially new therapeutic approaches. With respect to the heart, the newly introduced antianginal compound bepridil has probably the highest specifity for this organ. Voltage clamp measurements of isolated cardiac tissue have shown that in such preparations bepridil is an active CA, like classical compounds, such as verapamil or diltiazem. Besides its calcium antagonistic activity, bepridil also blocks the rapid-sodium channel which explains its antiarrhythmic potency of the class I type according to Vaughan Williams. In spite of its obvious activity at the level of cardiac calcium channels, bepridil, in therapeutic doses, has little or no effect on peripheral vascular smooth muscle [for reviews on bepridil, see ref. 27, 28]. Verapamil and gallopamil (D 600) act prodominantly on the heart, with a special affinity for the atrioventricular (A–V) node or the pacemaker cells, which is the basis for their antiarrhythmic potency. Apart from this well-known activity on cardiac tissues, verapamil is also a potent vasodilator, which is the basis of its potential use in the therapy of arterial hypertension and its antianginal activity. Diltiazem shows a similar profile although its vascular component may be somewhat stronger than that of verapamil. Its impairing influence on A–V conduction is possibly less than that of verapamil. As such it is intermediate between verapamil and nifedipine.

The *dihydropyridines*, of which nifedipine is the prototype, mainly cause relaxation of blood vessels, in particular of the coronary and peripheral arteries. Given by mouth in therapeutic doses, nifedipine does not display direct activity on cardiac tissues and the compound is not used as an antiarrhythmic agent. The mild and transient tachycardia caused by nifedipine reflects a reflex event mediated by the baroreceptor system [29]. Within the series of dihydropyridines, compounds with special effects on distinct vascular beds may be anticipated. Nimodipine, for instance, is being emphasized as a potential drug used for cerebral vasodilatation and it appears possibly useful in conditions of cerebral ischemia [30, 31]. Apart from cerebral vasodilatation, nimodipine is assumed to display anti-ischemic activity at the cellular level. Under physiological conditions, the diphenylalkylamines flunarizine or cinnarizine have little or no influence on the peripheral cardiovascular system [32] or on the brain. However, their protective activity on the *ischemic* brain is well-established and therapeutically appreciated [33]. A protective effect of these compounds in conditions of myocardial ischemia is the subject of present studies by our group. It should be realized that flunarizine and cinnarizine display little or no vasodilator activity, although they slightly, and probably nonspecifically, inhibit vasopressor effects of a variety of vasoconstrictors, including catecholamines, serotonin and angiotensin II [32]. Flunarizine can be shown to depress myocardial contractility under particular experimental conditions [34], although this effect is very weak and irrelevant during oral treatment with flunarizine.

It may be of interest to add that flunarizine is an effective antidote to digitalis poisoning in guinea pigs [35], probably because of its ability to impair the overloading of cardiac cells with calcium ions, as observed in the presence of toxic concentrations of a cardiac glycoside.

The obvious organ specificity of the various CAs so far developed may be explained by differences in calcium channel characteristics and density in the various organs and tissues. The complexity and diversity of calcium channels offer ample possibilities for a largely different pattern of calcium channels in different regions and tissues.

Pharmacological Background of Therapeutic Applications

Cardiovascular Disease

Cardiac Dysrhythmia. Both the automaticity of the cardiac pacemaker cells and the process of A–V conduction are dependent upon the influx of calcium ions through specific channels. Conversely, specific inhibition of the calcium inward flux by CAs is the basis of the antidysrhythmic effect of these drugs. In particular, verapamil and gallopamil are used as antidysrhythmic agents in cardiology in order to counteract supraventricular tachycardia. In view of the foregoing A–V blockade is a logical side effect of these CAs, reflecting the calcium dependency of A–V conduction. Diltiazem displays antidysrhythmic activity but it is not currently used as such [for a review of the antidysrhythmic activity of CAs, see ref. 36, 37].

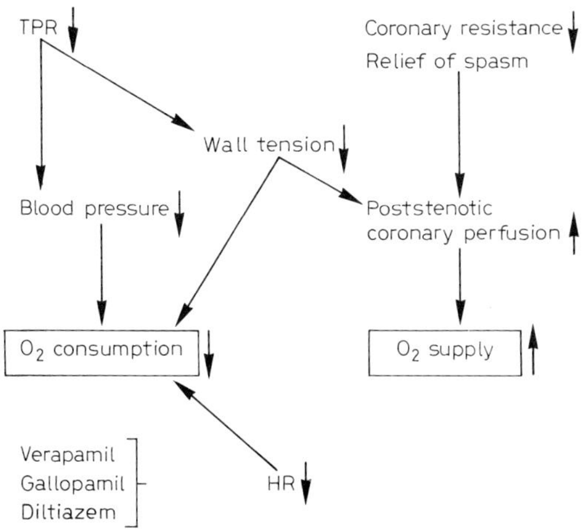

Fig. 4. Schematic presentation of the potential mechanism of CAs with respect to their beneficial effect in angina pectoris. The final result is an improvement of the imbalance between myocardial oxygen demand and supply. TPR = Total peripheral resistance; HR = heart rate.

Angina pectoris. Angina pectoris is a well-established indication for verapamil, the dihydropyridines and diltiazem. The mechanisms underlying this therapeutic benefit are manifold: relief of coronary spasm (when present); coronary vasodilatation *not* accompanied by a steal effect; a reduction in total peripheral resistance and, hence, cardiac afterload (only in the case of verapamil and diltiazem, *not* for the dihydropyridines); a decrease in cardiac frequency. An anti-ischemic mechanism at the cellular level may contribute to the antianginal effect of the CAs, although so far this remains uncertain. This combination of mechanisms leads to an improvement in the unfavorable balance O_2 supply to the heart/O_2 consumption which is characteristic of angina pectoris [for reviews on this subjects, see ref. 38, 39, and fig. 4].

Arterial Hypertension. Vasodilatation, predominantly in the arterial (arteriolar) vascular bed, is the main mechanism which satisfactorily explains the antihypertensive activity and hemodynamic properties of most of the dihydropyridine CAs (nifedipine, nitrendipine) and also that of verapamil. In modest doses, the vasodilatation is limited to the arteriolar vascular bed, leaving the capacitance vessels unaffected. Accordingly, total peripheral resistance is reduced. The dihydropyridines cause modest and transient reflex tachycardia as a response to peripheral vasodilatation. Verapamil, however, does not induce tachycardia but rather a tendency to bradycardia as a result of impaired A–V conduction [for reviews, see ref. 39, 41]

Other Applications of CA in Cardiovascular Disease. A transient relief of Raynaud's phenomenon by nifedipine is doubtless based upon vasodilatation in the digital arteries. Hypertrophic obstructive cardiomyopathy, pulmonary hypertension, congestive heart failure, the sick-sinus syndrome and acute digitalis intoxication are other potential applications of the CAs in cardiovascular disease. These highly specialized indications require further substantiation before they can be generally recommended and applied. These limited and somewhat uncertain applications of the CA will not be discussed in any detail in the present context [for reviews, see ref. 42, 43].

Calcium Antagonists in the Treatment of Neurological Disease

In contrast to the by now well-established position of the CAs in the management of several cardiovascular diseases, the attempted treatment of certain neurological disorders with these drugs constitutes a new field of investigation. This field, although most promising, is still full of uncertainty, both from the therapeutic and from the fundamental point of view. The results obtained so far strongly suggest that further exploration of the application of CAs in neurological disorders will be well worthwhile.

Compounds. So far, very few CAs have been studied appropriately as potential therapeutics in neurological disease. The diphenylalkylamine (fig. 1) *flunarizine* has been studied widely and has become one of the prototypes of CAs applied in neurological disorders. The efficacy of flunarizine in migraine prophylaxis and in the treatment of certain forms of vertigo appears to be generally accepted, although its usefulness in a variety of other disorders remains to be demonstrated. For a variety of reasons flunarizine appears to be preferable over cinnarizine [for reviews, see ref. 44–46].

Nimodipine is a dihydropyridine and thus related to nifedipine. This CA is claimed to be a vasodilator with a certain degree of specificity towards the cerebral vascular bed over the peripheral arterial system, although this specificity has been contested [47–49]. The vascular activity of nimodipine is much stronger than that of flunarizine which as such is not a vasodilator but rather causes a modest attenuation of the vasoconstrictor effects of endogenous factors like noradrenaline, angiotensin II, or serotonin. Apart from its vasodilator potency, nimodipine, like flunarizine, would be expected to display cytoprotective activity, especially in a condition of ischemia when the cellular calcium concentration becomes excessively high. Several therapeutic claims have been made for nimodipine, for instance, in subarachnoidal hemorrhage and cerebral ischemia. The potential activity of nimodipine in migraine is being investigated.

Mode of Action. In contrast to the application of CAs in cardiovascular disease, where the mode of action of the drugs is reasonably well understood, the mechanisms underlying the therapeutic application of CAs in neurology remain rather uncertain and subject to debate.

Improved perfusion of ischemic neuronal tissues appears to be a prominent effect of nimodipine, a CA with obvious vasodilator potency in the cerebral region. However, it should be realized that this vascular activity is probably not limited to the cerebral vessels, peripheral vasodilatation and concomitant hypotension should be anticipated as well. Furthermore, vasodilatation in the cerebral region may be the basis of steal effects which have indeed been claimed to occur.

Flunarizine, when given orally in low doses (5–10 mg daily), displays hardly any cardiovascular activity. Only *in vitro* has it been demonstrated that the vasoconstrictor activity of endogenous agonists, like serotonin, catecholamines, KCl and $PGF_{2\alpha}$, is blunted after pretreatment with flunarizine. It therefore seems likely, that the therapeutic activity of flunarizine is hardly due to a substantial degree by vasodilatation (or antivasoconstriction, as it is sometimes called) but rather explained on the basis of cytoprotection in a condition of ischemia, possibly by reducing the cellular overload with calcium, which is characteristic of the ischemic situation. A schematic representation of calcium overload blockade is shown in figure 5. Such a cytoprotective mechanism should also be claimed for nimodipine, since its vasodilator potency is probably not sufficient to explain all of its therapeutic effects [for reviews, see ref. 47–49]. An excessive rise in the concentration of extracellular potassium ions during cerebral ischemia is a generally recognized pathogenic process. This phenomenon is usually accompanied by a reduction in extracellular space which impairs the diffusion of oxygen and hence accelerates the development of ischemia. The elevated concentration of potassium in the

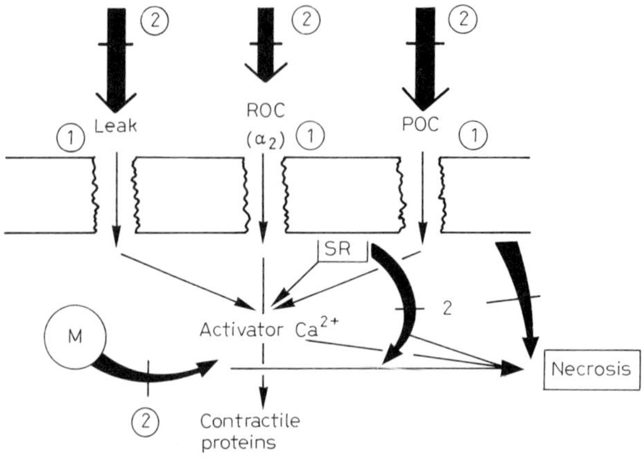

Fig. 5. Schematic presentation of the effect of CAs during ischemia. 1 = Inhibition of the calcium influx via different calcium channels by a calcium channel blocker under physiological conditions: leak channel; ROC, operated via α_2-adrenoceptors; POC, triggered by depolarization; 2 = Inhibition of the pathologically increased calcium influx under ischemic conditions, as indicated by the large black arrows (→). Apart from an excessive influx of calcium through the physiological channels there occurs an enhanced release of calcium from the inner side of the cell membrane (far right), as well as from subcellular pools like the sarcoplasmic reticulum (SR), mitochondria (M) or microsomes (M), resulting in a grossly increased cystolic calcium concentration, which significantly contributes to necrosis.

extracellular space reduces the transmembrane resting potentials of several types of cells. Owing to the potassium-induced depolarization, the influx of calcium ions into the cells is enhanced, causing vasoconstriction, release of neurotransmitters and calcium overload, leading to necrosis. Excessive calcium concentrations also cause increased stiffness of the erythrocytes which is counteracted by flunarizine. Accordingly, flunarizine-induced changes in the rheology of the blood may contribute to the drug's therapeutic activity in cerebral ischemia.

CAs of the flunarizine and nimodipine types are assumed to interfere with these extremely complex series of events, predominantly by inhibiting the cellular influx of calcium ions. It goes without saying that many details in this mechanism remain to be clarified.

Flunarizine has been claimed to be of therapeutic benefit as add-on medication in the treatment of certain forms of epilepsy [50], although this activity has been denied by others. The theoretical background is based upon the existence of a particular potential, derived from the brain's surface during an epileptic discharge, called the paroxysmal depolarizing shift. It spreads to deeper brain structures during more prolonged epileptic discharges. In experimental animals, it has been shown that the paroxysmal depolarizing shift is frequently preceded by a pathological influx of calcium ions, which may be sensitive to CAs. Experimental studies have shown that its extension is indeed strongly inhibited by verapamil and flunarizine. The relevance of these observations in human epilepsy remains to be demonstrated.

Finally, it should be emphasized that there is no conclusive evidence that the various therapeutic effects and claims of CAs in neurologic disorders are indeed directly related to calcium entry blockade, although this seems very likely.

Pharmacological Background of Adverse Reactions

The heterogeneity of the various CA implicates that the adverse reactions are also rather different for the various individual compounds. However, vasodilatation and its sequelae are very common side effects of all dihydropyridines, of verapamil and to some degree also of diltiazem. The background of the various side effects will be discussed here for the individual CAs or groups of CAs.

Dihydropyridines

Headache, flush, hypotension and reflex tachycardia are well-known side effects of the dihydropyridines (nifedipine, nitrendipine, nimodipine and related dihydropyridines). These adverse reactions all reflect vasodilation, based upon the inhibition of calcium influx through the specific channels. Peripheral edema, in particular in the legs, is a well-known side effect of nifedipine, which is *not* based upon the retention of sodium and fluid. This type of edema does not respond adequately to treatment with diuretics. It has been presumed that nifedipine would promote the opening of the sphincters of the capillary vessels and hence provoke the aforementioned edema.

Verapamil and Gallopamil

Similarly to the dihydropyridines, verapamil causes the side effects which are characteristic of vasodilatation, with the exception of reflex tachycardia. Verapamil rather causes bradycardia and impairment of A–V conduction, as a result of its depressant influence on calcium-influx-dependent processes in the sinus node and in the A–V node. Obstipation, a frequently reported side effect of verapamil, is caused by the relaxant effect of verapamil on colon movements, probably as a result of the inhibition of calcium fluxes in intestinal smooth muscle. As far as can be judged at present, the side effects of gallopamil are similar to those of verapamil.

Diltiazem

Headache and hypotension caused by diltiazem again reflect vasodilatation. Bradycardia, although occurring infrequently, is probably of the same origin as that described for verapamil. On the whole, the impression is that diltiazem is subjectively better tolerated than the dihydropyridines or verapamil, although hard data substantiating this impression are not available at present.

Flunarizine

Flunarizine, unlike the other CAs, exerts but little or no influence on the cardiovascular system and does not show the adverse reactions, like vasodilatation or cardiac depression, described above for other CAs. The most relevant adverse reactions to flunarizine are all based upon central nervous mechanisms.

Sedation is a fairly common side effect of flunarizine, probably resulting from a rather unspecific mechanism. Depressions caused by flunarizine have sometimes been observed, although usually in patients suffering from migraine. The association migraine/depression is not uncommon and it is therefore difficult to judge the role of flunarizine as such in this matter.

Extrapyramidal side effects attributed to flunarizine have been reported in rare cases and virtually only in the elderly. The mechanism of this adverse reaction so far remains unknown and is difficult to explain on a molecular basis, since flunarizine has no relevant affinity for dopaminergic receptors.

Finally, flunarizine, when applied for prolonged periods, sometimes causes a slight increase in body weight, which is usually transient. The mechanism of this effect is unknown. It should be recalled that several drugs with obvious central nervous activity, like benzodiazepines, neuroleptics and antidepressants cause a similar gain in body weight. It seems likely that the effect of flunarizine is not particularly specific.

Conclusions

The CAs have obtained a generally recognized position in the treatment of cardiovascular disease, in particular of angina pectoris, arterial hypertension and certain cardiac dysrhythmias. Most of their therapeutic activities, including their adverse reactions in the cardiovascular field, are understood reasonably well and as such provide an intellectually satisfactory picture of these drugs in basic pharmacology.

The application of certain CAs in neurology has opened a new field of enormous potential interest, still characterized by many uncertainties and speculations, both from the fundamental/mechanistic and from the practical/therapeutic points of view.

Migraine and vertigo have become routine indications for flunarizine, whereas the therapeutic claims of nimodipine in subarachnoid hemorrhage and in acute cerebral ischemia appear to be rather firm and well substantiated. Flunarizine might develop as an antiepileptic drug of a novel type and with a relatively low incidence of adverse reactions. Owing to the availability of flunarizine for *intravenous* application, it is now possible to study its potential benefit in acute stroke. Trials designed to study this issue are proceeding at present. Most of the results strongly indicate that a field of new therapeutic possibilities has been opened. The enormous investment in further clinical and fundamental research required to elucidate the many uncertainties appears to be well worthwhile. It will be the task of basic pharmacologists to unravel the mechanisms underlying the therapeutic efficacy and side effects of the CAs in neurological disorders.

References

1 Fleckenstein, A.: Specific pharmacology of calcium in the myocardium cardiac pacemaker and vascular smooth muscle. Annu. Rev. Pharmacol. Toxicol. *17:* 149–166 (1977).
2 Vanhoutte, P.M.; Bohr, D.F.: Calcium entry blockers and the cardiovascular system. Fed. Proc. *40:* 2851 (1981).
3 Nayler, W.G.; Poole-Wilson, P.: Calcium antagonists: definition and mode of action. Basic Res. Cardiol. *76:* 1–15 (1981).
4 Nayler, W.G.: Calcium antagonists. Eur. Heart J. *1:* 225–237 (1980).
5 Paoletti, R.; Vanhoutte, P.M.: International Symposium on Calcium Antagonists. Pharmacology and Clinical Research, New York City 1987.
6 Borgers, M.: Personal communication.
7 Van Zwieten, P.A.: Calcium antagonists – terminology, classification and comparison. Arzneimittel-Forsch. *35:* 298–301 (1985).

8 Van Zwieten, P.A.: Differentiation of calcium entry blockers into calcium channel blockers and calcium overload blockers. Eur. Neurol. *25:* suppl. 1, pp. 57–67 (1986).

9 Mannhold, R.; Rodenkirchen, R.; Bayer, R.: Qualitative and quantitative structure-activity relationships of specific CA-antagonists. Progr. Pharmacol. *5:* 25–52 (1982).

10 Rodenkirchen, R.; Bayer, R.; Mannhold, R.: Specific and nonspecific analysis of cardiodepressive drugs. Progr. Pharmacol. *5:* 9–23 (1982).

11 Rodenkirchen, R.; Bayer, R.; Steiner, R.; Bossert, F.; Meyer, H.; Möller, E.: Structure-activity studies on nifedipine in isolated cardiac muscle. Arch. Pharmacol. *310:* 69–78 (1979).

12 Spedding, M.: Calcium antagonist subgroups. Trends pharmacol. Sci. *6:* 109–114 (1985).

13 Glossmann, H.; Ferry, D.R.; Lubbecke, F.; Mewes, R.; Hofmann, F.: Calcium channels: direct identification with radioligand binding studies. Trends pharmacol. Sci. *3:* 431–437 (1982).

14 Schwartz, A.; Triggle, D.J.: Cellular action of calcium channel blocking drugs. A. Rev. Med. *35:* 325–339 (1984).

15 Glossmann, H.; Goll, A.; Rombusch, M.; Ferry, D.R.: Molecular pharmacology of $C\alpha^{2+}$-channels: receptor binding studies; in Betz, Deck, Hoffmeister, Nimodipine; pharmacological and clinical properties, pp. 57–76 (Schattauer, Stuttgart 1985).

16 Spedding, M.: Three types of $C\alpha^{2+}$-channels explain discrepancies. Trends pharmacol. Sci. *8:* 115–117 (1987).

17 Godfraind, T.; Miller, R.C.; Socrates Lima, J.: Selective α_2- and α_1-adrenoceptor agonist-induced contractions and ^{45}Ca-fluxes in the rat isolated aorta. Br. J. Pharmacol. *77:* 597–604 (1982).

18 Fleckenstein, A.: Calcium antagonism in heart and smooth muscle. Experimental facts and therapeutic prospects (Wiley, New York 1983).

19 Peters, T.: Calcium in physiological and pathophysiological cell function. Eur. Neurol. *25:* suppl. 1, pp. 27–44 (1986).

20 Bou, J.; Llenas, J.; Massingham, R.: Calcium entry blocking drugs, “calcium antagonists” and vascular smooth muscle function. J. auton. Pharmacol. *3:* 219–232 (1983).

21 Triggle, D.J.; Swamy, V.C.: Calcium antagonists. Some chemical-pharmacologic aspects. Circulation Res. *52:* suppl. I, pp. 17–28 (1983).

22 Peters, T.: Calcium-Antagonismus. Ein einheitlicher Begriff für uneinheitliche Wirkmechanismen. Med. Klin. *78:* 368–375 (1983).

23 Van Meel, J.C.A.; De Jonge, A.; Wilffert, B.; Kalkman, H.O.; Timmermans, P.B.M.W.M.; Van Zwieten, P.A.: Organic and inorganic calcium antagonists reduce vasoconstriction in vivo mediated by postsynaptic α_2-adrenoceptors. Arch. Pharmacol. *316:* 288–293 (1981).

24 Van Meel, J.C.A.; Towart, R.; Timmermans, P.B.M.W.M.; Van Zwieten, P.A.: Correlation between the inhibitory activities of calcium antagonists on vascular smooth muscle constriction in vitro after depolarisation and in vivo after α_2-adrenoceptor stimulation. Arch. Pharmacol. *322:* 34–37 (1983).

25 Van Zwieten, P.A.; Van Meel, J.C.A.; Timmermans, P.B.M.W.M.: Pharmacological basis of the therapeutic effect of calcium entry blockers. Interaction with vascular α-adrenoceptors. Hypertension *5:* suppl. II, pp. II-8–II-17 (1983).

26 Van Zwieten, P.A.; Timmermans, P.B.M.W.M.; Van Heiningen, P.N.M.: Receptor subtypes involved in the action of calcium antagonists. J. Hypertension 5 (suppl. 4): 521–528 (1987).

27 Labrid, C.; Grosset, A.; Dureng, G.; Mironneau, J.; Duchêne-Marullaz, P.: Some membrane interactions with bepridil, a new antianginal agent. J. Pharmac. exp. Ther. *211:* 546–554 (1979).
28 Marshall, R.J.; Winslow, E.; Lamar, J.C.; Apoil, E.: Bepridil; in Scriabine, New drugs annual: cardiovascular drugs, vol. 2, pp. 157–176 (Raven Press, New York 1984).
29 Lederballe Pedersen, O.; Christensen, N.J.; Ramsch, K.D.: Comparison of acute effects of nifedipine in normotensive and hypertensive man. J. cardiovasc. Pharmacol. *2:* 357–361 (1980).
30 Gelmers, H.J.: The effects of nimodipine on the clinical course of patients with acute ischemic stroke. Acta neurol. scand. *69:* 232–239 (1984).
31 Paci, A.; Ottaviono, B.; Trenta, A.; Iannone, G.; Acuto, G.; Carosi, M.: Treatment of acute ischemic cerebral infarction with nimodipine (BAY e 9736); in pp. 473–478. Nimodipine. Pharmacological and clinical properties. Proc. 1st. Int. Nimotop-Symposium, München 1984 (Schattauer, Stuttgart 1985).
32 Van Nueten, J.M.; Vanhoutte, P.M.: Calcium entry blockers and vasospasm; in Vanhoutte, Leusen, Vasodilation pp. 459–468 (Raven Press, New York 1982).
33 Wauquier, A.; Edmonds, H.C.; Van den Broek, W.A.E.; Melis, W.; Van Loon, J.: Effects of flunarizine on recovery of cerebral function following cardiac arrest. Neuropsychobiology *13:* 147–151 (1985).
34 Boddeke, H.W.G.M.; Heynis, J.B.; Jonkman, F.A.M.; Van Zwieten, P.A.: Characterization of calcium entry blockers in isolated guinea-pig hearts; cardiodepressant activity of flunarizine. Br. J. Pharmacol. *89:* 615P (1986).
35 Jonkman, F.A.M.; Boddeke, H.W.G.M.; Van Zwieten, P.A.: Protective activity of calcium entry blockers against ouabain intoxication in anaesthetized guinea pigs. J. cardiovasc. Pharmacol. *8:* 1009–1013 (1986).
36 Godfraind, T.; Herman, A.; Wellens, H.: Calcium entry blockers in cardiovascular and cerebral dysfunctions, pp. 1–25 (Nijhoff, Den Haag 1984).
37 Packer, M.; Frishman, W.H.: Calcium channel antagonists in cardiovascular disease, pp. 225–269 (Appleton-Century-Crofts, Norwalk 1984).
38 Braunwald, E.: Mechanism of action of calcium channel blocking agents. New Engl. J. Med. *307:* 1618–1627 (1982).
39 Van Zwieten, P.A.: Drug targets in unstable angina; in Hugenholtz, Goldman, Unstable angina, current concepts and management, pp. 151–157 (Schattauer, Stuttgart 1985).
40 Van Zwieten, P.A.; Timmermanns, P.B.M.W.M.: Pharmacological basis of the antihypertensive action of calcium entry blockers. J. cardiovasc. Pharmacol. *7:* S11–S17 (1985).
41 Van Zwieten, P.A.; Schönbaum, E.: Calcium antagonists and blood vessels. Progr. Pharmacol. *5:* 1–142 (1983).
42 Bussey, H.F.; Talbert, R.L.: Promising uses of calcium channel blocking agents. Pharmacotherapy *4:* 137–143 (1984).
43 Gray Ellrodt, A.; Singh, B.N.: Clinical applications of slow channel blocking compounds. Clin. pharmacol. Ther. *23:* 1–43 (1983).
44 Van Nueten, J.M.; Van Beek, J.; Janssen, P.A.J.: Effect of flunarizine on calcium-induced response of peripheral vascular smooth muscle. Archs int. Pharmacodyn. Thér. *232:* 42–52 (1978).

45 Van Nueten, J. M.; Janssen, P. A. J.: Comparative study of the effects of flunarizine and cinnarizine on smooth muscles and cardiac tissues. Archs int. Pharmacodyn. Thér. *204:* 37–55 (1973).

46 Hartmann, A.; Van Zwieten, P. A.: Pharmacological aspects and neurological potential of calcium entry blockers. 13th Wld. Congr. Neurol., Hamburg 1985. Eur. Neurol. *25:* suppl. 1, pp. 1–126 (1986).

47 Betz, E.; Deck, K.; Hoffmeister, F.: Nimodipine. Pharmacological and clinical properties. Proc. 1st Int. Nimotop-Symposium, München 1984 (Schattauer, Stuttgart 1985).

48 Tanaka, K.; Gotoh, F.; Muranratsu, F.; et al.: Effects of nimodipine (BAY e 9736) on cerebral circulation in cats. Arzneimittel-Forsch. *26:* 2172–2176 (1980).

49 Kazda, S. R.: Nimodipine: a new calcium antagonistic drug with a preferential cerebrovascular action. Acta neurochir. *63:* 259–265 (1982).

50 Overweg, J.; Ashton, D.; De Beukelaar, F.; Binnie, C. D.; Wauquier, A.; Van Wieringen, A.: Add-on therapy in epilepsy with calcium entry blockers. Eur. J. Neurol. *25:* 93–101 (1986).

P. A. van Zwieten, MD, PhD, Divisions of Pharmacotherapy and Cardiology, Academic Medical Centre, University of Amsterdam, Meibergdreef 15, NL–1105 AZ Amsterdam (The Netherlands)

Van Zwieten PA (ed): Clinical Aspects of Calcium Entry Blockers.
Prog Basic Clin Pharmacol. Basel, Karger, 1989, vol 2, pp 21–66

Hemodynamic Actions of Calcium Entry Blockers

H. A. J. Struyker-Boudier, J. G. De Mey, J. F. M. Smits, H. M. N. W. Nievelstein[1]

Department of Pharmacology, University of Limburg, Maastricht, The Netherlands

Introduction

Calcium entry blockers (CEBs) are now widely used in the treatment of patients with different cardiovascular diseases, e.g., ischemic heart disease, heart failure, Raynaud's phenomenon, pulmonary and arterial hypertension. Their therapeutic effectiveness in these disease states is based upon a range of hemodynamic actions, including effects on blood vessels, the heart, the kidneys as well as different neural and endocrine mechanisms involved in cardiovascular control. The degree to which each of these mechanisms is affected may differ per subgroup of CEBs. The pharmacological basis for the classification of the different subgroups of CEBs was dealt with in the preceding article. The purpose of the present contribution is to review the hemodynamic actions of CEBs as a basis of understanding their clinical effects in the treatment of cardiovascular diseases. This review is based upon a critical selection out of close to 1,000 papers published between 1980 and 1987 on the hemodynamic effects of CEBs. For a discussion of some of the older studies, the reader is referred to several earlier reviews [Fleckenstein, 1977, 1983; Godfraind, 1981; Flaim and Zelis, 1982; Cauvin et al., 1983; Nayler and Horowitz, 1983]. In addition, recent issues of the journal *Drugs* contain reviews on individual CEBs, e.g., diltiazem [Chaffman and Brogden, 1985], nitrendipine [Goa and Sorkin, 1987] and nicardipine [Sorkin and Clissold, 1987].

[1] The authors express their thanks to Dr. J. Urquhart for critical discussion and to Els Geurts and Mia Hogenboom for preparation of the manuscript.

The Cardiovascular System and Its Control Mechanisms

The ultimate hemodynamic effect of a drug used in the treatment of cardiovascular diseases is the result of (a) the primary pharmacological effect of the drug, and (b) the reactions of the cardiovascular system and its control mechanisms to the perturbation induced by the drug [Struyker-Boudier, 1980, 1984]. This implies potentially important differences between the in vitro or acute in vivo effects of a cardiovascular drug and its long-term actions in an experimental animal or human. A clear example is that of β-adrenoceptor blockers. The predominant in vitro or acute in vivo effect of β-blockers is a depression of cardiac function with relatively few vascular effects. Prolonged treatment of hypertensive animals or patients leads to an initial increase and later reduction of vascular resistance [Van Baak et al., 1985]. This differential vascular effect is thought to be due to an early activation and later adaptation of the sinoaortic baroreceptor reflex [Struyker-Boudier et al., 1979; Smits et al., 1982; Van Baak et al., 1985].

We have reviewed the role of cardiovascular control mechanisms in the effects of antihypertensive drugs elsewhere [Struyker-Boudier, 1984]. Briefly, these mechanisms can be distinguished on the basis of their physiological nature and time characteristics. The most rapid form of control is exerted by nervous reflexes, of which the sinoaortic baroreflex is the most important. These nervous reflexes are powerful and rapidly acting negative feedback mechanisms. Blood pressure is continuously monitored by sensory afferents. Via the central nervous system these reflexes influence the activity of the autonomic nervous system, and thereby the heart and vasculature. The sinoaortic baroreflex is a potent short-term controller of blood pressure. It is much less active in the long-term, since it adapts its activity to the prevailing blood pressure level [Guyton, 1986].

The intermediate term form of cardiovascular control is exerted by endocrine mechanisms, the most important of which are the renin-angiotensin-aldosterone system, the antidiuretic hormone and probably the recently described atrial natriuretic factor. A detailed discussion of these endocrine systems is beyond the scope of this contribution. As control mechanisms, they act relatively rapidly and are able to influence different segments of the cardiovascular system, e.g., blood vessels, the heart, but also renal and capillary sites of fluid exchange. In this way, they provide both short- and long-term forms of cardiovascular control.

Long-term control of the cardiovascular system is exerted by the kidneys on the one hand and by the structural design of the cardiovascular system on the other hand. The renal control of blood pressure is described in detail in Guyton's [1986] textbook of physiology. The contribution of the structural design of the cardiovascular system in its long-term functioning has not yet been fully determined. However, it now seems that adaptive changes in the myocardium and blood vessel wall may partly determine the maintenance of abnormal hemodynamics in different cardiovascular disease states, such as essential hypertension and congestive heart failure [Lever, 1986].

Vascular Effects of Calcium Entry Blockers

Effects on Isolated Blood Vessels

There is little doubt that CEBs are able to dilate precontracted isolated blood vessels. The vasodilator effect of CEBs is due to a direct effect on vascular smooth muscle and does not depend on an interference with the release of noradrenaline from sympathetic nerve terminals. The vasodilatation is most pronounced when the contraction is induced by an increased transmembrane influx of calcium ions. Van Breemen et al. [1979] and Bolton [1979] proposed that calcium influx in vascular smooth muscle can be stimulated by different pathways. The pathway stimulated by K^+ depolarization was referred to as the potential-operated channel (POC) and that by receptor occupation as the receptor-operated channel (ROC). Recent studies indicate that even at the level of a single smooth muscle cell more than one type of POC or ROC may be found [for a review, see Hurwitz, 1986]. In addition, Hwa and Bevan [1986] have proposed the existence of a stretch-dependent calcium entry pathway.

Originally, it was assumed that POCs are more sensitive to the action of CEBs than ROCs [Cauvin et al., 1983]. Indeed, in large arteries, such as the rat or rabbit aorta, K^+-induced contractions are much more readily inhibited by CEBs than noradrenaline-induced contractions [Schumann et al., 1975; Godfraind and Miller, 1982; Hof et al., 1982, 1984a; Towart, 1982; Godfraind, 1983; Kazda et al., 1983; Kannan et al., 1986]. Furthermore, the efficacy of inhibition of K^+-induced contractions by CEBs is markedly constant for different arterial preparations [Cauvin et al., 1983].

However, the higher sensitivity of POCs is not a general rule for all types of blood vessels. The aorta cannot be regarded as a vessel of great importance in the control of vascular resistance. Therefore, several recent studies have used smaller, resistance-sized blood vessels. These studies indicate a similar sensitivity of POCs and ROCs to the actions of CEBs. In comparison to the rabbit aorta, the isolated rabbit mesenteric artery is much more sensitive to the blockade of noradrenaline-induced contractions by diltiazem [Van Breemen et al., 1981], nifedipine [Schumann et al., 1975], cinnarazine [Brockaert and Godfraind, 1979] and nisoldipine [Saida and Van Breemen, 1983]. Diltiazem causes an equal degree of inhibition of K^+- and noradrenaline-induced contractions of rabbit basilar, central ear and mesenteric arteries [Bevan, 1982; Bevan et al., 1986] and rat femoral arteries [Fujimoto et al., 1986]. Bijak et al. [1986] have recently compared the effects of isradipine on different types of isolated vessels contracted either by high potassium or by histamine. Potassium-induced contractions were inhibited to an almost equal degree in each vessel type (IC_{50} : 3–20 × 10^{-11} *M*). Histamine-induced contractions were not affected in the aorta or renal artery ($IC_{50} > 10^{-6}$ *M*), but were inhibited in the coronary, basilar and central ear arteries (IC_{50} : 2–3 × 10^{-9} *M*).

Julou and Freslon [1986] determined the IC_{50} values of four different CEBs on K^+-versus noradrenaline-induced contractions of rat mesenteric resistance vessels. They found comparable IC_{50} values for each CEB, e.g., verapamil (1.3 vs. 2 × 10^{-7} *M*), diltiazem (3 vs. 4.5 × 10^{-7} *M*), nifedipine (4.5 vs. 9 × 10^{-9} *M*) and nicardipine (3 vs. 7 × 10^{-9} *M*). Cauvin et al. [1987] made a similar observation for isradipine and nisoldipine in the mesenteric resistance vessels of spontaneously hypertensive rats.

Further evidence for the relative lack of selectivity of CEBs on K^+ and noradrenaline in resistance vessels comes from the work of Sybertz et al. [1983]. These authors used a perfused rat hindquarter preparation to investigate the interaction of CEBs with different vasoconstrictive stimuli. Nifedipine, diltiazem and verapamil were only slightly more potent against potassium- than noradrenaline-induced vasoconstriction.

Several possible explanations exist for the difference between large and small arteries in POC versus ROC sensitivity for CEBs. The first is that noradrenaline may depolarize smooth muscle cells from resistance-sized vessels, and thereby act partly through POCs. There is indeed some evidence from the work of Mulvany's group [Nyborg and Mulvany, 1984; Nyborg et al., 1985] to support this view. Thus, at the level of small arteries, noradrenaline promotes the entry of Ca^{2+} into smooth muscle cells by activating both types of calcium-gating mechanisms.

The second explanation is related to differences between mechanisms of electromechanical and pharmacomechanical coupling. Whereas electromechanical coupling mediated by POCs relies primarily upon transmembrane influx of calcium, pharmacomechanical coupling depends upon both influx of external calcium and release from internal stores [Bohr, 1963; Van Breemen et al., 1972]. Since most studies on the effects of CEBs on isolated aortae used relatively large concentrations of noradrenaline (10^{-7} to 10^{-4} *M*), the contraction may have depended largely upon the release of internal calcium. Cauvin et al. [1983] have suggested that the high sensitivity of resistance vessels to the CEB inhibition of noradrenaline-induced constriction is due to the lack of an intracellular calcium component.

A third possible explanation for the variability in the ROC sensitivity between different blood vessels to CEBs has been the implication of a differential effect of CEBs on noradrenaline-induced contractions via α_1- or α_2-adrenoceptors.

Calcium Entry Blockers and Vascular α-Adrenoceptors

A few years ago, Van Meel et al. [1981] proposed that vasoconstriction mediated by α_2-adrenoceptors is dependent on the influx of extracellular calcium, whereas vasoconstriction mediated by α_1-adrenoceptors is not. The evidence for this hypothesis was based upon studies in the pithed rat, in which the pressor effects of α_2-agonists were more readily inhibited by CEBs than were the pressor effects of α_1-agonists [Van Meel et al., 1981; Timmermans et al., 1983a, 1984; Gerold and Haeusler, 1983; Jie et al., 1985]. Similar observations were made in the ganglion-blocked rabbit [Van Meel et al., 1982], pithed cat [Timmermans et al., 1983b], perfused rat hindquarters [Van Meel et al., 1983], ganglion-blocked rat [Lew and Angus, 1985], isolated rat aorta [Godfraind et al., 1982], perfused dog hindlimb [Saeed et al., 1983], isolated rat superior mesenteric artery [Kannan and Seip, 1986], perfused rat tail artery [Hicks et al., 1985], and forearm of hypertensive patients [Jie et al., 1987]. A recent whole-animal study by Lappe et al. [1985] supports the view that CEBs preferentially block α_2-adrenoceptor-mediated vasoconstriction by endogenous catecholamines. They studied the effects of nisoldipine on vascular reactivity to humoral and neural vasoconstrictor stimuli in conscious, intact rats. Whereas nisoldipine attenuated the vasoconstrictor actions of exogenously applied noradrenaline and adrenaline, it did not affect vasoconstriction induced by posterior hypothalamic or splanchnic nerve stimulation [Lappe et al., 1985].

However, there is no general agreement on the preferential sensitivity of CEBs for α_2-adrenoceptor-mediated vasoconstriction. De Mey and

Vanhoutte [1981], in a study on the isolated splenic artery and saphenous vein of dogs, demonstrated a higher sensitivity of α_1-compared to α_2-adrenoceptors to the inhibitory effect of verapamil. In later studies, similar observations were made on isolated canine coronary arteries [Rimele et al., 1983; Müller-Schweinitzer, 1983], perfused canine hindlimb [Jandhyala, 1985], perfused rat renal arteries [Holck and Gerold, 1985], perfused rat hindquarters [Pedrinelli and Tarazi, 1985b], anesthetized cats [Alabaster and Solca, 1985], perfused feline mesenteric vascular bed [Armstead et al., 1987], canine saphenous vein [Jayakody et al., 1987b], spinal dogs [Morita et al., 1985], human digital arteries [Stevens and Moulds, 1986], intact conscious dogs, rhesus monkeys [Woodman et al., 1986] and normotensive men [Pasanini et al., 1985].

Holck and Gerold [1985] questioned the possible reasons for this discrepancy. They used a pithed rat preparation in which they measured the effects of drugs or spinal cord stimulation on vascular resistance rather than blood pressure, as was done in the original studies by Van Meel et al. [1981]. They observed that the vasoconstrictor effects of α_1-adrenoceptor stimulation were much more readily blocked by CEBs than the pressor actions. In pithed rats an increase in cardiac output plays a major role in contributing to the pressor response following both spinal stimulation and the addition of α-adrenoceptor agonists [Holck and Gerold, 1985; Vollmer and Fluharty, 1987; Nievelstein et al., unpubl. observations]. Apart from β-adrenoceptor stimulation, the rise in cardiac output may be caused by a direct stimulation of cardiac α-adrenoceptors or by constriction of capacitance vessels. Since the relative contribution of α_1- and α_2-adrenoceptors to each of these effects is still unknown, inferences on the basis of changes in blood pressure may be misleading. In fact, in a recent study Pedrinelli and Tarazi [1985b] documented a clear distinction between the pressor effects of systemic or local (autoperfused hindquarter) actions of α-adrenoceptor agonists in pithed rats.

A second explanation for the discrepancies in the studies on the relative contributions of α_1- and α_2-adrenoceptors in the effects of CEBs is of a molecular pharmacological nature. Several authors have shown the existence of a large receptor reserve for the postsynaptic α_1-adrenoceptor with a much smaller receptor reserve for the postsynaptic α_2-adrenoceptor [Hamilton et al., 1983; Ruffolo and Yaden, 1984]. When α_1-adrenoceptor reserve is reduced by pretreatment with the irreversible α-blocker phenoxybenzamine, the effects of full α-agonists are much more sensitive to inhibition by CEBs in the pithed rat [Ruffolo et al., 1984; Bou and

Massingham, 1986; Timmermans et al., 1985; Jim et al., 1986], canine saphenous vein [Cooke et al., 1985], and ganglion-blocked rats [Pedrinelli and Tarazi, 1985a]. Similarly, the pressor effects of partial α_1-adrenoceptor agonists in the pithed rat are much more sensitive to antagonism by CEBs than are full agonists [Timmermans et al., 1983a; Jim et al., 1985, 1986; Chiu et al., 1986].

The third explanation for the differences observed in various studies on the inhibition of α_1- and α_2-adrenoceptor effects by CEBs comes from observations made by Scarborough and Carrier [1984] and McGrath and O'Brien [1987]. These investigators found that differences in the time course of effects exerted by α-adrenoceptor agonists play a role in the differential effects of CEBs. In isolated rat aortae α-adrenoceptor-induced contraction can be separated into a fast and slow component. Nifedipine blocks the slow, but not the fast component [Scarborough and Carrier, 1984]. In intact rats, bolus injections of α-agonists cause a rapid rise in blood pressure which reaches a peak within 30 s. After an initial rapid decline a second response is observed as a shoulder on the decline of the first phase [McGrath and O'Brien, 1987]. This second, slower component is much more susceptible to the inhibitory effect of nifedipine. Both in the in vitro and the in vivo study, it was impossible to distinguish the two phases on the basis of an α_1- versus α_2-adrenoceptor-mediated effect. Upon infusion in intact rats, the initial rapid pressor component was eliminated. Nifedipine in that situation inhibited the effects of α_1- and α_2-agonists to a similar degree. McGrath and O'Brien [1987] suggested that in intact animals hemodynamic rather than receptor-mediated events are responsible for the differential sensitivity of CEBs. The acute hemodynamic effect of α-agonists would be a CEB-insensitive arteriolar constriction, whereas the slow- and long-term effects involve additional postcapillary and cardiac effects. Further in vivo microcirculatory studies are needed to prove this hypothesis.

In conclusion, the concept that differential ROC sensitivity of CEBs can be explained by supposing that vasoconstriction mediated by α_2-adrenoceptor stimulation is dependent on the influx of extracellular calcium, whereas vasoconstriction mediated by α_1-adrenoceptors is not, should now be regarded as too simplistic. The choice of animal model, type of agonist (full or partial), the elimination of receptor reserve and the schedule of agonist administration all determine the outcome of the effects of CEBs. In addition, it has not been established to what degree interference of CEBs with catecholamine-induced α-adrenoceptor stimulation contributes to the in vivo vasodilatory effect of these drugs in patients or intact experimental animals. Recent studies show that CEBs interfere with the pressor effects of additional endogenous substances, such as serotonin [Müller-Schweinitzer and Neuman, 1983], angiotensin II [Hof, 1984a, 1985] and neuropeptide Y [Mabe et al., 1985; Pernow and Lundberg, 1986; Pernow et al., 1987]. Furthermore, for the in vivo situation the general and very sensitive interaction of CEBs with POCs may be of primary importance.

Does the Endothelium Play a Role in the Vascular Effects of Calcium Entry Blockers?

In 1980 Furchgott and Zazwadzki demonstrated that acetylcholine relaxes the rabbit aorta via the release of a relaxing factor from the endothelium (EDRF). In the meantime the mechanisms governing the release of EDRF have been investigated intensively. The Ca^{2+} ion seems of particular importance in the release process of EDRF [Rapoport et al., 1984; Long and Stone, 1985; Peach et al., 1987]. Singer and Peach [1982] showed that nifedipine and verapamil block the EDRF-dependent relaxation of rabbit aorta. Winquist et al. [1985] could not confirm this observation. In a recent study, Jayakody et al. [1987a] were not able either to show an inhibitory effect of diltiazem or nicardipine on the endothelium-dependent relaxation to acetylcholine. The direct relaxant effect by CEBs of isolated blood vessels contracted with potassium is not influenced by removal of the endothelium [Williams et al., 1987]. Therefore, up to now it does not seem that the endothelium plays an important role in the acute vascular effects of CEBs.

In Vivo Vascular Effects of Calcium Entry Blockers

In the previous section the evidence for CEB-induced vascular relaxation was reviewed primarily from the point of view of the effects on isolated vascular preparations. It is clear that contracted large (conduit) arteries, small (resistance) arteries as well as veins can be dilated by CEBs, although there may be quantitative differences in the degree of dilatation of various vessel types.

Effects on vascular tone should ultimately be reflected in hemodynamic changes in the whole organism, in particular in changes in vascular resistance.

Effects on Total Peripheral Vascular Resistance. A first approach to the in vivo assessment of drug effects on vascular resistance is the determination of their action on overall systemic hemodynamics. Following this approach, effects on blood pressure, heart rate and cardiac output are usually measured experimentally. Subsequently, changes in stroke volume and total peripheral vascular resistance (TPVR) are calculated using standard formulae.

Table I summarizes the systemic hemodynamic effects of a series of CEBs in normal humans and in essential hypertensive or coronary heart disease patients. Studies on patients suffering from congestive heart failure, or patients using other drugs simultaneously with a CEB have not been included in this table. All CEBs cause a fall in mean arterial pressure (MAP) and TPVR. This effect occurs both after acute and during long-term administration of CEBs. Thus, these studies clearly confirm vasodilatation as a primary mechanism of the antihypertensive action of CEBs.

The effects of CEBs on cardiac hemodynamic variables are less consistent. Upon acute administration, the dihydropyridines in general cause a rise in heart rate (HR) and cardiac output (CO). These effects are most likely caused by an activation of the sino-aortic baroreflex as a consequence of the drop in blood pressure [Nievelstein et al., 1985; see below section entitled 'Nervous Control of the Circulation, p. 42]. Upon chronic administration of dihydropyridines, the reflex effects on CO and HR disappear to a large extent or entirely. This is probably due to a gradual diminution of baroreflex influences subsequent to an adaptation of sino-aortic baroreceptors to the prevailing arterial blood pressure [Nievelstein et al., 1985; Kleinbloesem et al., 1987].

The effects of dihydropyridines on stroke volume, or other parameters for cardiac contractility, is even less consistent. A few authors (table I) find a significant, although only small, increase. Both a reduced afterload and a reflex increase in sympathetic activity may be involved in the increase in stroke volume. On the other hand, a reduced preload – as a consequence of venodilatation or a reduction of intravascular volume [Fagard et al., 1987] – counteracts this effect. Thus, the influence of CEBs on stroke volume may vary with the balance between pre- and afterload effects.

The hemodynamic actions of verapamil and diltiazem in humans differ from those of the dihydropyridines. Although these CEBs also reduce MAP by a decrease in TPVR, they cause no tachycardia and only small changes in CO. The rise in CO is not found consistently. In fact, the significant increase following diltiazem reported by Amodeo et al. [1986] was not reported in an earlier paper from the same group [Frohlich, 1985]. The earlier paper indicated a statistically insignificant rise from 3.1 to 4.3

Table I. Summary of the effects of acute (A) or chronic administration of different CEBs on mean arterial pressure (MAP), heart rate (HR), cardiac output (CO), stroke volume (SV) and total peripheral vascular resistance (TPVR) in normal (N), and essential hypertensive (EH) or coronary heart disease (CHD) patients

Drug	dose	Duration	Patients	MAP	HR	CO	SV	TPVR	Authors
Verapamil	0.15 mg/kg + 1–2 mg/min	A, i.v.	EH	–	0	+	+	–	Cody, 1987
Verapamil	80–160 mg t.i.d.	1 week	EH	–	0	0	+	–	Cody, 1987
Verapamil	40–80 mg t.i.d.	1 year	EH	–	–	0	+	–	Lund-Johansen, 1983a
Diltiazem	0.2 mg/kg	A, i.v.	EH	–	0	0	0	–	Safar et al., 1983
Diltiazem	0.06 mg/kg	A, i.v.	EH	–	0	+	0	–	Amodeo et al., 1986
Diltiazem	0.5 mg/min	A, i.v.	EH	–	0	0	0	–	Aoki et al., 1983
Diltiazem	0.25 mg/kg + 1.4 μg/kg/min	A, i.v.	CHD	–	0	0	+	–	Joyal et al., 1985
Diltiazem	0.5 mg/kg	A, i.v.	CHD	–	0	0	0	–	Legrand et al., 1984
Diltiazem	0.5 mg/kg	A, i.v.	CHD	–	–	0	+	–	Silke et al., 1987
Diltiazem	120–180 mg b.i.d.	4 weeks	EH	–	0	0	0	–	Amodeo et al., 1986
Nifedipine	20 mg	A, oral	CHD	–	+	+		–	Majid et al., 1982
Nifedipine	20 mg	A, oral	CHD	–	0	0	0	–	Lehmann et al., 1983
Nifedipine	10 mg	A, oral	EH	–	+	+	0	–	Kiowski et al., 1983
Nifedipine	10 mg b. or q.i.d.	3 weeks	EH	–	0	+	0	–	Olivari et al., 1979
Nifedipine	20–40 mg b.i.d.	1 year	EH	–	0	0	0	–	Lund-Johansen, 1983b
Nifedipine	20 mg t.i.d.	6 weeks	EH	–	0	0	0	–	Kiowski et al., 1983
Nisoldipine	1.5 μg/kg	A, i.v.	CHD	–	+	+	0	–	Vogt et al., 1980
Nisoldipine	8 μg/kg	A, i.v.	CHD	–	+	+	+	–	Silke et al., 1985
Nitrendipine	20 mg	A, oral	EH	–	+	+	0	–	Ventura et al., 1983
Nitrendipine	20 mg	A, oral	EH	–	+	0	0	–	Nannan et al., 1984
Nitrendipine	1 mg	A, i.v.	EH	–	+	+	0	–	Rutsch and Schmutzler, 1984
Nitrendipine	20–40 mg b.i.d.	2 weeks	EH	–	0	0	+	–	Fouad et al., 1984
Nicardipine	10 mg	A, i.v.	CHD	–	+	+	+	–	Silke et al., 1984
Nicardipine	10 mg	A, i.v.	CHD	–	+	+		–	Rousseau et al., 1984
Nicardipine	2 mg + 50 μg/min	A, i.v.	CHD	–	+	+	+	–	Lambert et al., 1985
Nicardipine	2 mg + 0.2–1 mg/min	A, i.v.	EH	–	+	+	0	–	Ryman et al., 1987
Felodipine	15 mg	A, oral	N	–	+	+	0	–	Agner et al., 1985
Felodipine	0.1 mg/kg	A, oral	EH	–	+	+	+	–	Andersson et al., 1985
Felodipine	0.075 mg/kg	A, oral	EH	–	+	+	+	–	Fagard et al., 1987

– = Decrease; + = increase; 0 = no significant change.

liters/min/m^2 for cardiac index following the administration of diltiazem to 7 essential hypertensive patients. The later paper [Amodeo et al., 1986] contained the results of 9 patients with essential hypertension and 1 with primary aldosteronism. It is not clear whether this comprised an entirely new group of patients, but the rise from 3.3 to 4.1 liters/min/m^2 was significant in this case.

There are several possible explanations for the relative lack of cardiac effects of verapamil and diltiazem when compared to acutely administered dihydropyridines. First, verapamil and diltiazem have direct effects on the heart, which counteract the reflex activation of the cardiac variables. Second, verapamil and diltiazem may interfere more strongly than dihydropyridines with sino-aortic reflex control of the circulation. Both aspects of their action will be discussed in more detail in later sections.

Systemic hemodynamic studies in experimental animals confirm the observations in man. Table II summarizes the results of a number of studies on the hemodynamic effects of acute administration of different CEBs in various animal models. All CEBs lower blood pressure by a reduction in TPVR. The dihydropyridines almost invariably increase CO and HR. The exceptions with respect to the increase in HR are usually found in anesthetized animals. Since anesthetics suppress the nervous reflex control of the circulation, it is not surprising that the baroreflex-mediated increase in HR, normally observed in intact organisms, is less prominent in anesthetized animals.

As in man, verapamil and diltiazem cause little reflex rise in cardiac variables. HR is even sometimes reduced, probably as a consequence of the direct cardiac effects of these substances.

Hof et al. [1986] have recently studied the hemodynamic effects of the stereoisomers of isradipine. In vitro studies have indicated that the (+)-enantiomer is at least 100-fold more active as a CEB. In the in vivo experiment this enantiomer also much more readily reduced MAP. However, both enantiomers increased CO and reduced vascular resistance. The (−)-enantiomer in particular reduced coronary vascular resistance. Hof et al. [1986] interpret these data as an indication that calcium channels not only discriminate between stereoisomers, but also respond differently to them. Further research with stereoisomers may reveal more about the role of different calcium channels in the control of vascular resistance.

Only very few studies have thus far been aimed at an analysis of the long-term hemodynamic effects of CEBs in animal models for hypertension. A 12-week treatment of spontaneously hypertensive rats (SHR) with

Table II. Summary of the effects of different CEBs on mean arterial pressure (MAP), heart rate (HR), cardiac output (CO), stroke volume (SV) and total peripheral vascular resistance (TPVR) in experimental animals

Drug	dose	Species[1]	Anesth.[2]	MAP	HR	CO	SV	TPVR	Authors
Verapamil	1 mg/kg i. v.	cat	C/U	–	–	0		–	Hof, 1983
Verapamil	1 mg/kg i. v.	SHR	C	–	0	0	0	–	Nievelstein et al., 1985
Diltiazem	1 mg/kg i. v.	cat	C/U	–	–	+		–	Hof, 1983
Diltiazem	10 mg/kg/h i. v.	NR	C	–	0	0	0	–	Flaim and Zelis, 1982b
Diltiazem	10 mg/kg/h i. v.	SHR	C	–	–	+	+	–	Flaim et al., 1986
Nifedipine	0.4 mg/kg oral	dog	C	–	+	+	0	–	Gross et al., 1979
Nifedipine	1.5 mg/kg/h i. v.	NR	C	–	+	0	0	–	Kanda and Flaim, 1984
Nifedipine	0.3 mg/kg i. v.	SHR	C	–	+	+	0	–	Nievelstein et al., 1985
Nisoldipine	1–10 μg/kg i. v.	dog	F	–	+	+	0	–	Kazda et al., 1980
Nisoldipine	1.6 μg/kg/min i. v.	NR	C	–	+	0	0	–	Drexler et al., 1985
Nisoldipine	0.3 mg/kg oral	NR	C	–	+	+	0	–	Drexler and Just, 1986
Nisoldipine	4–25 μg/kg i. v.	cat	P	–		+		–	Armstead et al., 1987
Nisoldipine	1 μg/kg/min i. v.	pig	A/M	–	+	0	0	–	Duncker et al., 1986a
Nitrendipine	1–10 μg/kg i. v.	dog	F	–	+	+	0	–	Stoepel et al., 1981
Nitrendipine	8 μg/kg/min i. v.	dog	C	–		+		–	Knight et al., 1985
Nitrendipine	10 mg/kg oral	SHR	C	–	0	0	0	–	Pegram et al., 1984
Nicardipine	100 μg/kg i. v.	cat	C/U	–	0	+		–	Hof, 1983
Nicardipine	1–10 μg/kg i. v.	dog	P	–	0	+	+	–	Takanake et al., 1985
Felodipine	100 nmol/kg i. v.	RH rabbit	C	–	+	+	0	–	Bolt and Saxena, 1984
Felodipine	500 nmol/kg i. v.	SHR	C	–	+	+	+	–	Nordlander, 1985
Nimodipine	1.25 μg/kg/min i.v.	pig	A/M	–	–	–	0	–	Duncker et al., 1986b
Darodipine	50 μg/kg i. v.	cat	C/U	–	–	+		–	Hof, 1983
Darodipine	0.3 mg/kg i. v.	SHR	C	–	+	+	0	–	Nievelstein et al., 1985
(±) Isradipine	1–10 μg/kg i. v.	cat	C/U	–	–	+		–	Hof et al., 1984b
(±) Isradipine	3 μg/kg i. v.	dog	C/U	–	0	+		–	Hof et al., 1984b
(+) Isradipine	1–10 μg/kg i. v.	cat	C/U	–	–	+		–	Hof et al., 1986
(–) Isradipine	30–300 μg/kg i. v.	cat	C/U	0	0	+		–	Hof et al., 1986
Flunarazine	3 mg/kg i. v.	dog	L	–	+	+	0	–	Kubo et al., 1984

– = Decrease; + = increase; 0 = no significant change.

[1] Species: SHR = spontaneously hypertensive rat; NR = normotensive rat; RH = renal hypertensive.

[2] Anesthesia: C = conscious; C/U = chloralose/urethane; A/M = azaperone/metomidate; P = pentobarbital; F = fentanyl; L = lofentanyl.

felodipine lowered blood pressure and TPVR [Nordlander, 1985]. After this treatment period HR had not changed, but CO and stroke volume were significantly increased [Nordlander et al., 1985]. These data suggest a decrease of resistance to venous return of blood to the heart. A 3-week treatment of SHR with nitrendipine also lowered blood pressure and TPVR, without significant changes in cardiac variables [Kobrin et al., 1984]. A recent study on a 3-week treatment of SHR with diltiazem gave inconsistent results. Although blood pressure was slightly reduced, no other hemodynamic variable changed significantly during diltiazem treatment [Natsume et al., 1985].

In summary, studies in intact animals as well as in humans confirm that the primary mode of antihypertensive action of all CEBs is peripheral vasodilatation.

Effects on Regional Vascular Resistances. Data reviewed in the previous section show that CEBs invariably reduce blood pressure by a decrease in TPVR. It can be questioned whether the decrease in TPVR is brought about by an equal reduction of resistance in different vascular beds. Originally, many CEBs were believed to act particularly on coronary vascular resistance [Fleckenstein, 1977, 1983]. Later, claims were made for other forms of regional selectivity, e.g., nimodipine was proposed to act preferentially on cerebral vessels [Towart and Kazda, 1979].

Most of the original claims on preferential vascular activity of CEBs were based on studies on isolated blood vessels or on in vivo studies using merely one vascular bed. With the recent introduction of more sophisticated regional flow measurement techniques, e.g., radioactive microspheres or chronically implanted Doppler and electromagnetic flow probes, it has become possible to study the in vivo effects of CEBs on simultaneously determined parallel vascular resistances. Table III summarizes the results of a number of such studies upon acute administration of different CEBs.

The data contained in table III reveal some surprising results. First, there is a marked difference in regional effects on vascular resistance for the various CEBs. They invariably reduce coronary and skeletal muscle vascular resistance to a strong degree. The only two exceptions with respect to the effects of skeletal muscle come from studies in which nisoldipine [Drexler et al., 1986] or nitrendipine [Pegram et al., 1984] were administered orally in rats. Oral administration in the rat notoriously leads to unpredictable drug bioavailability and in both studies only small effects on

Table III. Summary of the effects of different CEBs on regional vascular resistances in experimental animals.

Drug	dose	Species[1]	Anesth.[2]	Method[3]	Skel. muscle	Skin	Brain	Heart	Kidneys	Intest.	Authors
Verapamil	1 mg/kg i.v.	cat	C/U	M	–	0	–	--	0	–	Hof, 1983
Verapamil	3 mg/kg i.v.	SHR	C	D	--				0	0	Nievelstein et al., 1985
Diltiazem	1 mg/kg i.v.	cat	C/U	M	–	0	–	--	–	–	Hof, 1983
Diltiazem	10 mg/kg/h i.v.	NR	C	M	--	0	–	--	–	--	Flaim and Zelis, 1982
Diltiazem	10 mg/kg/h i.v.	SHR	C	M	--	0	--	--	–	–	Flaim et al., 1986
Nifedipine	1.5 mg/kg/h i.v.	NR	C	M	--	–	0	--	–	--	Kanda and Flaim, 1984
Nifedipine	0.3–1 mg/kg i.v.	SHR	C	D	--				0	0	Nievelstein et al., 1985
Nifedipine	0.1 mg/kg i.v.	SHR	P	M	--		--	--	–	–	Reed and Tuma, 1986
Nisoldipine	1.6 µg/kg/min i.v.	NR	C	M	--	0	–	–	–	–	Drexler et al., 1985
Nisoldipine	0.3 mg/kg oral	NR	C	M	0	0	–	--	–	–	Drexler et al., 1986
Nisoldipine	1 µg/kg/min i.v.	pig	A/M	M	--	–	–	--	0	–	Duncker et al., 1986a
Nitrendipine	1–10 µg/kg i.v.	dog	F	EM	--	–	–	--		–	Stoepel et al., 1981
Nitrendipine	8 µg/kg/min i.v.	dog	C	EM					–	--	Knight et al., 1985

blood pressure were found, in spite of the use of massive doses. In addition, all CEBs affect vascular resistance in the skin very little. In fact, almost all authors report a redistribution of blood flow from the skin to skeletal muscle and the myocardium. The effects on renal vascular resistance are usually relatively small and run in parallel to the effects on blood pressure. This indicates that the reduction of resistance in this bed may, at least partly, be the consequence of an autoregulatory response to the drop in perfusion pressure. The net effect is a maintenance of renal blood flow. Finally, the effects on brain and (gastro)intestinal resistance are somewhat more variable. These two vascular beds, like the kidney, are known to contain relatively powerful mechanisms to control their flows via autoregulation. However, in a number of studies additional dilatory effects of CEBs were found in these vascular beds.

Table III. (cont.)

Drug	dose	Species[1]	Anesth.[2]	Method[3]	Skel. muscle	Skin	Brain	Heart	Kidneys	Intest.	Authors
Nitrendipine	10 mg/kg oral	SHR	C	M	0	0	0	–	0	0	Pegram et al., 1984
Nicardipine	0.1 mg/kg i.v.	cat	C/U	M	– –	0	– –	– –	–	–	Hof, 1983
Nicardipine	1–10 μg/kg i.v.	dog	P	EM	–		– –	– –	0	–	Takanaka et al., 1985
Felodipine	0.5 μmol/kg i.v.	SHR	C	M	– –		–	– –	–	– –	Nordlander, 1985
Felodipine	0.1 μmol/kg i.v.	RH rabbit	C	M	– –	0	– –	– –	–	– –	Bolt and Saxena, 1984
Nimodipine	1.25 μg/kg/min i.v.	pig	A/M	M	– –	0	–	– –	0	–	Duncker et al., 1986b
Darodipine	50μg/kg i.v.	cat	C/U	M	– –	0	– –	– –	–	0	Hof, 1983
Darodipine	0.1–1 mg/kg i.v.	SHR	C	D	– –				–	0	Nievelstein et al., 1985
(±) Isradipine	10 μg/kg i.v.	cat	C/U	M	– –		– –	– –	–	–	Hof et al., 1984b

0 = no significant effect; – = a reduction in the order of magnitude as the fall in TPVR; – – = a reduction stronger than that of TPVR

[1] Species: SHR = spontaneously hypertensive rat; NR = normotensive rat; RH = renal hypertensive.

[2] Anesthesia: C = conscious; C/U = chloralose/urethane; A/M = azaperone/metomidate; P = pentobarbital; F = fentanyl.

[3] Method of flow measurement: M = microspheres; D = Doppler flow probes; EM = electromagnetic flow probes.

A second surprising result from the data summarized in table III is that – in contrast to the marked heterogeneity for different vascular beds – the various CEBs act similarly, even in different species. This observation suggests that vascular regional specificity is not a result of differences in the structural properties of the tissue target sites ('receptors') for CEBs in different vascular beds, as was suggested before [Hof, 1984b]. Rather, the data point to the direction of heterogeneity in the properties of calcium-gating mechanisms beyond the 'receptor' level or of differences in local vascular control in various vascular beds. There is thus far only very limited evidence for heterogeneity in the calcium-gating mechanisms. Therefore, explanations for tissue selectivity of CEBs have been sought for mostly in the direction of local vascular control.

Bolt and Saxena [1984] have pointed out that the strong vasodilatory effects of CEBs in the heart and brain may be due to metabolic autoregulatory effects, in addition to the direct actions of CEBs on the vasculature. However, the same authors indicated that there is little evidence for increased tissue metabolism following the administration of CEBs. In fact, the afterload reduction may even decrease myocardial oxygen consumption.

Nordlander [1985] and Knight et al. [1985] suggested that the degree of vasodilatation induced by CEBs may be related to resting vascular tone. Thus, skeletal muscle which has a high resting vascular tone would be dilated much more than the renal vascular bed with its low basal tone. Hwa and Bevan [1986] doubt the importance of myogenic tone in determining the vascular effect of CEBs, since dihydropyridine CEBs do not influence the stretch-dependent myogenic tone in different vascular preparations [Hwa and Bevan, 1986].

An additional explanation for the regional selectivity of CEBs comes from the work of Nievelstein et al. [1985]. These authors showed that the preferential skeletal muscle dilatation following administration of nifedipine to intact SHR is changed into a nonselective dilatation in sino-aortic-baroreflex-denervated SHR (fig. 1). These data suggest that the effects of sympathetic activation due to baroreflex stimulation in intact rats are more effectively counteracted by nifedipine in skeletal muscle than in the kidney or intestinal vascular bed. Studies are in progress now to establish whether this is the consequence of a selective interference of nifedipine at a central site in the baroreflex loop or of differential effects of nifedipine on noradrenaline (α-)receptor populations in various vascular beds. A recent study in the pithed rat indeed indicates that – in contrast to skeletal muscle which has a mixed α_1/α_2-receptor population – adrenergic vasoconstriction in the kidney and gastrointestinal system is almost exclusively due to α_1-adrenoceptor stimulation [Hiley and Thomas, 1987].

Regional hemodynamic effects of CEBs in humans have been studied only in a limited number of vascular beds for obvious technical reasons. The effects on coronary and renal vascular bed will be discussed in subsequent sections. The most frequently studied other vascular bed is the forearm. A significant decrease in forearm vascular resistance was reported for various CEBs, e.g., verapamil [Cody, 1987], nifedipine [Kiowski et al., 1983; Ferguson and Dorsey, 1985], nicardipine [Thuillez et al., 1984; Levenson et al., 1985], diltiazem [Finch and Johnston, 1985], nitrendipine [Takeshita and Mark, 1980; Levenson et al., 1984], and felodipine [Mace et al., 1985; Sluiter et al., 1985; Agner et al., 1985]. It is impossible, however, to conclude that regional selectivity of CEBs exists in normal or hypertensive subjects on the basis of such studies.

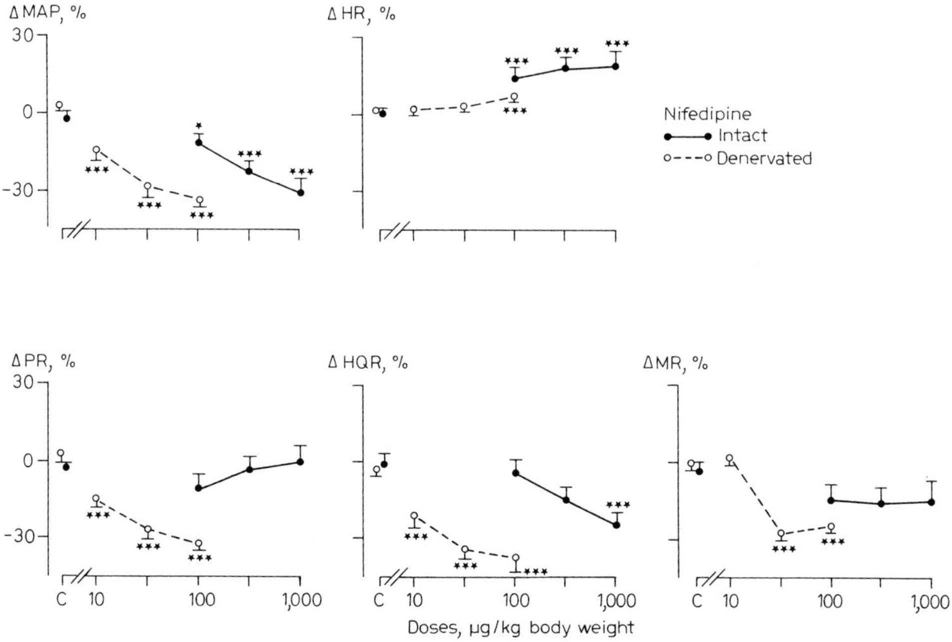

Fig. 1. Hemodynamic effects of different doses of nifedipine in conscious intact and sino-aortic-baroreflex-denervated SHR. MAP=Mean arterial pressure; HR=heart rate; RR=renal resistance; HQR=hindquarter resistance; MR=mesenteric resistance.

In summary, the predominant hemodynamic cause of CEB-induced reduction of blood pressure is a decrease in vascular resistance. It seems that CEBs act in two fundamentally different modes, depending upon whether or not they are engaging the sympathetic nervous system via the baroreceptor reflexes. In the absence of sympathetic nerve action CEBs cause generalized vasodilatation. This effect is negated in the presence of sympathetic reflex activity. In that case CEBs predominantly dilate the coronary and skeletal muscle vascular bed.

Effects on Venous Function. Calcium-dependent vascular contractions have been described for both arterial and venous preparations. Thus, CEBs can be expected to dilate arterial as well as venous vessels. This has indeed been observed in studies on isolated blood vessels, although a higher concentration of the CEBs was usually needed to dilate precontracted venous preparations [cf., Cauvin et al., 1983].

The in vivo effects of CEBs on veins are of a more complex nature. When infused intra-arterially into a human forearm preparation in doses that had no systemic effects, verapamil was about 20 times more effective in dilating arterioles than veins [Robinson et al., 1979]. Upon intravenous or oral administration, the direct venodilatory effect of CEBs is opposed by baroreflex-mediated sympathetic venoconstriction [Segstro et al., 1986]. In an elegant study in anesthetized dogs, Ogilvie [1985] showed that nifedipine – in contrast to the converting enzyme inhibitor captopril – increases venous return of blood to the heart. Furthermore, this author indicated that the effects on peripheral flow distribution appear to be a more important factor in determining total venous return than the effects on the venous vasculature. The increase in venous return may explain the frequently observed rise in stroke volume following the administration of dihydropyridines.

Effects on the Microcirculation. Most studies on the vascular effects of CEBs have focussed on large arteries or veins. However, much of the actual control of vascular resistance is exerted at the level of the microcirculation. Recently, the effects of CEBs on isolated resistance-sized vessels have been determined (see above, section entitled 'Effects on Isolated Blood Vessels', p. 23). Very few data are available on the in vivo effects of CEBs on the microcirculation. Altura et al. [1980] used intravital microscopy to study the effects of verapamil on rat mesenteric, pial and cremaster muscle arterioles as well as muscular venules. Local application of verapamil to cremaster muscle arterioles and venules caused vasodilatation and increased capillary perfusion. However, verapamil did not affect the mesenteric and pial vasculature [Altura et al., 1980]. In microscopy experiments in the cat, Haws and Heistad [1984] also found little effect of nimodipine on resting pial arteriolar diameters. However, nimodipine was able to inhibit the vasoconstrictor responses to sympathetic nerve stimulation or hypocapnia [Haws and Heistad, 1984].

Direct microscopic visualization of the rat left ventricular epicardial microcirculation has shown the potent coronary vasodilator effect of nimodipine [McDonagh et al., 1984].

Finally, Colantuoni et al. [1984] used the hamster skin fold window preparation, which allows the observation of the microcirculation without anesthesia. In that preparation, verapamil modestly dilated arterial vessels, with the most pronounced effects on the smallest arterioles. Clearly, more research is needed before any definite conclusions can be drawn on potential microcirculatory sites and mechanisms of action of CEBs.

Cardiac Effects of Calcium Entry Blockers

The original discovery of CEBs as a unique group of drugs was made in relation to their cardiac effects. The 'calcium antagonist' properties were originally based on the observation that the ability of these drugs, in particular verapamil, to uncouple excitation-contraction coupling in heart muscle could be reversed by the addition of Ca^{2+} [Fleckenstein, 1977, 1983]. In the meantime it has become evident that the cardiac actions of CEBs involve more than a negative inotropic effect, but also include effects on cardiac rhythm, metabolism and coronary vascular tone. These aspects will each be discussed briefly. Detailed reviews of the cardiac effects of CEBs are given in a monograph by Fleckenstein [1983] and recent review articles [Nayler and Horowitz, 1983; Singh, 1986].

Effects on Myocardial Contractility

Cardiac muscle excitation-contraction involves the activation of a calcium inward current. Since this current is slowly activated when compared to the fast Na^{+} inward current it is sometimes referred to as the 'slow-Ca^{2+} current'. There are approximately 10,000 functional Ca^{2+} channels per cardiac muscle cell [Reuter, 1983]. The various groups of CEBs bind differently to the cellular region of, or close to, the Ca^{2+} channels. This is probably the basis for the different efficacies of various CEBs to influence the myocardium. In particular, dihydropyridines are much less active on the POCs of the myocardium than on vascular smooth muscle POCs [Chiu, 1986]. For verapamil and diltiazem more comparable values are obtained for these two tissue types. Ljung [1985] compared the sensitivity of rat vascular versus myocardial actions of CEBs and found equal sensitivities for verapamil and D-600. Diltiazem was 7-fold more active in the vascular preparation, whereas nifedipine and felodipine were respectively 14 and 118 times more active in the vascular preparation [Ljung, 1985]. This differential sensitivity has important pharmacodynamic consequences. In intact organisms dihydropyridine derivatives often exert positive rather than negative inotropic effects. In this case, reflex compensatory mechanisms via the sino-aortic baroreceptors exceed any potential direct negative inotropic effect. In addition, the afterload reduction may contribute to the positive inotropic effect [Campbell et al., 1985]. After the administration of verapamil or diltiazem, direct negative inotropic and indirect compensatory positive inotropic effects usually cancel each other (table I).

Effects on Nodal Tissue

Certain CEBs, in particular verapamil and diltiazem, slow atrioventricular nodal conduction and sinus pacemaker activity. The dihydropyridine and piperazine derivatives are much less effective in this respect. In intact organisms these agents may even increase heart rate. The tachycardia is not due to a direct effect, but rather to a baroreflex-mediated alteration of sympathetic and parasympathetic activity [De Langen et al., 1984]. The verapamil- and diltiazem-induced direct effects on nodal tissue may be due to a phenomenon referred to as 'use dependence' [Nayler and Horowitz, 1983; Nayler and Dillon, 1986]. This phenomenon refers to a situation in which drug potency increases as the frequency of membrane depolarization, and hence of channel opening, is enhanced. It would be limited to drugs that can only influence the functioning of the channel in its active or open state [Nayler and Dillon, 1986]. Verapamil and probably diltiazem, but not the dihydropyridines, provoke this phenomenon [Rowland et al., 1979]. The particular influence of verapamil and diltiazem on A-V nodal conduction has given these drugs a place in the treatment of certain cardiac arrhythmias [Singh, 1986].

Effects on Coronary Vascular Tone

In previous paragraphs the marked effects of CEBs on coronary vascular resistance have been discussed. In isolated preparations of coronary artery smooth muscle derived from laboratory animals and man, CEBs inhibit spontaneous contractile activity as well as contractions induced by depolarization or agonist substances. In intact animals, the administration of CEBs leads to a redistribution of blood flow, favoring the coronary vascular bed (see above). The coronary vasodilatory effect of CEBs has also been established in numerous studies using isolated perfused hearts, anesthetized open- or closed-chest dogs, cats and pigs, as well as angina pectoris patients undergoing cardiac catheterization. In such patients, direct observation of the coronary vasculature indicated that the intravenous administration of diltiazem increases coronary artery diameter at rest [Bonzel et al., 1984] and during exercise [Hossack et al., 1984].

Recently, there has been some discussion with respect to the effects of different vasodilators on various segments of the coronary vasculature, e.g., the small-resistance vessels, the large epicardial arteries and the collateral vessels [Schwartz and Bache, 1987]. CEBs dilate both large and small vessels, as opposed to nitroglycerin, which acts predominantly on the large coronary arteries [Brown et al., 1981a, b) and dipyridamole which has a relatively strong effect on the small resistance-sized vessels [Fam and McGregor, 1968]. Results of studies on the effects of CEBs on coronary collateral vessel formation or native collateral flow have been contradictory [Schwartz and Bache, 1987]. It is not yet clear whether differential actions on these various segments of the coronary vasculature have important therapeutic implications.

The reasons for the particular sensitivity of the coronary vascular bed to the vasodilator effects of CEBs remain to be established. One interesting possibility comes from the recent work of Hayashi et al. [1986]. These authors showed that nifedipine inhibits the coronary vasoconstrictor effects of neuropeptide Y in anesthetized, closed-chest dogs. Neuropeptide Y is found in the heart of many species, including man, and may be involved as a cotransmitter of the sympathetic nervous system in coronary vasoconstriction [Lundberg et al., 1983; Clarke et al., 1987]. A recent study by Pernow and Lundberg [1986] indicates that neuropeptide Y also constricts human skeletal muscle arteries. This effect is again blocked by nifedipine [Pernow and Lundberg, 1986]. It now looks as if there is a marked heterogeneity of autonomic nervous system peptide cotransmitters in different vascular beds [Burnstock, 1986]. A selective interference of CEBs with one of these neuropeptides could be the basis of their preferential sensitivity for the coronary and the skeletal muscle vascular bed.

Effects on Cardiac Metabolism and Damage due to Ischemia

One of the earliest areas of clinical application of CEBs was in the treatment of patients with angina pectoris. CEBs can influence the detrimental effects of myocardial ischemia in different ways. First, the coronary vasodilator effect enhances myocardial oxygen supply. Second, the reduced afterload as well as – especially in the case of verapamil and diltiazem – negative inotropic effect decrease cardiac work and oxygen demand. Third, CEBs cause a redistribution of blood flow in the ischemic myocardium with a relative increase to the subendocardial region [Bache and Tockman, 1982; Matzusaki et al., 1984]. Finally, CEBs may provide protection against the normally damaging effects of intracellular calcium overload. The elevation of intracellular calcium is associated with a reduction of adenosine triphosphate and other high-energy phosphate compounds [Fleckenstein, 1983]. Ultimately this process leads to irreversible damage of myocardial tissue, and thus to loss of mechanical and electrical activity of the heart.

The clinical consequences of these favorable actions of CEBs have been clearly established for patients with angina pectoris. However, the clinical outcome of cardioprotection by CEBs against the consequences of myocardial infarction remain to be established.

Effects of Calcium Entry Blockers on Cardiovascular Control Mechanisms

Nervous Control of the Circulation

The most important nervous control mechanism of the circulation is the sinoaortic baroreflex. This reflex is activated by a reduction in the pressure perfusing the sinoaortic arterial region. Several studies indicate that changes in flow through this area – as is the case when cardiac output is reduced – can also activate the sinoaortic baroreflex [Struyker-Boudier et al., 1979; Charlton and Baertschi, 1982]. Since CEBs induce a fall in arterial blood pressure an early stimulation of the sinoaortic reflex can be expected. Such activation of the sinoaortic baroreflex is expressed as an increase in sympathetic tone and rise in plasma noradrenaline levels, as was shown in man after the acute administration of nifedipine [Murphy et al., 1982; Kiowski et al., 1983; Saito et al., 1986], nitrendipine [Ventura et al., 1983; Lopez et al., 1984; Luft et al., 1985], nicardipine [Reyman et al., 1987], and felodipine [Sluiter et al., 1985; Fagard et al., 1987], but not after verapamil [Spies et al., 1978; Leonetti et al., 1982] and diltiazem [Amodeo et al., 1986]. Furthermore, the increase in sympathetic tone contributes to acute positive inotropic and chronotropic effects of the dihydropyridine CEBs. The relative lack of positive inotropic and chronotropic actions of verapamil and diltiazem is due to the direct cardiac effects of these agents. In addition, a more pronounced inhibitory effect of these agents on the sinoaortic baroreflex has been proposed [Heesch et al., 1983; Nakaya et al., 1983].

Most studies on the effects of CEBs on the baroreceptor reflex have used intravenous administration of the drugs. Although several investigators report an effect of CEBs on baroreflex sensitivity [Nakaya et al., 1983; Warltier et al., 1984; Taylor and Kowalski, 1984; Ferguson and Dorsey, 1985], it is impossible to conclude from these studies the site of action of CEBs within the reflex pathway.

Heesch et al. [1983] studied the effects of nifedipine and verapamil in a vascularly isolated canine carotid sinus preparation, while measuring carotid sinus nerve activity. Their studies indicate that nifedipine increases and verapamil decreases the sensitivity of the carotid sinus baroreceptors. Kunze et al. [1986] recently could not confirm this observation in an isolated perfused carotid sinus preparation in the cat or aortic arch preparation in the rat. In fact, they found that verapamil, diltiazem and a number of dihydropyridines only affect baroreceptor discharge at very high, therapeutically irrelevant, doses. They therefore conclude that a possible effect of CEBs on the baroreflex originates from a site other than the afferent nerves.

An alternative site of interaction with the baroreflex pathway may be the central nervous system. Several groups have recently proposed a central site of cardiovascular effects of CEBs [Higuchi et al., 1985, Iyoda et al., 1985; Laurent et al., 1987]. These studies were based upon the direct administration of verapamil [Higuchi et al., 1985], diltiazem (Higuchi et al., 1985; Iyoda et al., 1985] or nifedipine [Higuchi et al., 1985; Laurent et al., 1987] to the rat brain. Preferential sites of hypotensive action seem to be the nucleus tractus solitarii, the dorsal surface of the brain stem at the obex and the anterior hypothalamic/preoptic region. The indicated sites of action are known to contain noradrenergic synapses involved in the regulation of blood pressure [Struyker-Boudier et al., 1975]. Therefore, it cannot be excluded that the central actions of CEBs are mediated by changes in central noradrenergic neurotransmission.

Kleinbloesem et al. [1987] have suggested recently that the rate of increase in plasma concentration of nifedipine determines its hemodynamic effects in humans due to different forms of baroreceptor activation. Thus, these authors showed that a rapid rise in plasma concentration of nifedipine causes an immediate rise in HR, with no change in arterial blood pressure, whereas a gradual (5–7 h) asymptotic rise to the same steady-state plasma concentration has no effect on HR, but causes an asymptotic decline in diastolic blood pressure. This observation has important theoretical as well as therapeutic implications. The reasons for the differences in the baroreflex activation are not clear, but may be related to different hemodynamic actions of the two regimens. The study of Kleinbloesem et al. [1987] shows that a drop in blood pressure may not be responsible for the early baroreflex activation during the rapid rise in plasma nifedipine level. The rapid administration may, however, cause a much more drastic afterload reduction. Additional studies with direct cardiac and arterial hemodynamic measurements could help to analyze further this interesting phenomenon.

Little attention has thus far been given to the long-term effects of CEBs on nervous control of the circulation. Plasma noradrenaline levels are not elevated during chronic treatment of hypertensive patients with nifedipine [Kiowski et al., 1983] or diltiazem [Amodeo et al., 1986], although Pedrinelli et al. [1986] found a small increase in plasma catecholamine level after 2 weeks of a nitrendipine therapy. Littler's group has shown a resetting without alteration in sensitivity of the sinoaortic baroreflex during chronic treatment of hypertensive humans with felodipine [Smith et al., 1986] or nicardipine [Littler et al., 1986]. During long-term nifedipine therapy the sinoaortic baroreflex was reset to a lower pressure with a slight increase in sensitivity [McLeay et al., 1983].

Endocrine Control of the Circulation

Several endocrine systems can influence the circulation. The most important of these are the renin-angiotensin-aldosterone system, the antidiuretic hormone and possibly the more recently described atrial natriuretic factors.

The reflex rise in sympathetic nervous activity causes an increase in plasma renin following the acute administration of nifedipine [Lederballe-Pedersen et al., 1979; Hiramatsu et al., 1982; Kiowski et al., 1983; Saito et al., 1986], nicardipine [Van Schayk et al., 1984], or felodipine [Sluiter et al., 1985; Leonetti et al., 1986; Fagard et al., 1987] in man. Also in conscious rabbits, dihydropyridine CEBs cause an acute rise in plasma renin [Hof et al., 1987]. Acute administration of verapamil or diltiazem to hypertensive subjects does not lead to a rise in plasma renin [Leonetti et al., 1982; Guthrie et al., 1983; Aoki et al., 1983; Magometschnigg et al., 1984].

During chronic treatment of hypertensive patients with CEBs, plasma renin activity usually remains unchanged. This has been observed for different CEBs, including nifedipine [Kiowski et al., 1983; McLeay et al., 1983], nicardipine [Coruzzi et al., 1985; Young et al., 1985; Chaignon et al., 1986], nitrendipine [Simon and Snyder, 1984], diltiazem [Amodeo et al., 1986], and verapamil [Leonetti et al., 1980; Lewis, 1980; Muiesan et al., 1982]. Interestingly, during treatment of hypertensive patients with nitrendipine for a period of 1–2 weeks, plasma renin remained increased, whereas aldosterone levels were not affected [Fouad et al., 1984; Luft et al., 1985; Pedrinelli et al., 1986]. This has led to the suggestion that CEBs may dissociate plasma renin activity from that of aldosterone. There is indeed some evidence from studies on isolated calf adrenal glomerulosa cells that

CEBs inhibit angiotensin-II-induced aldosterone secretion [Kojima et al., 1984].

There are only few data on the effects of CEBs on antidiuretic hormone or atrial natriuretic factors. Furthermore, the interpretation of such data would be difficult, since the release of these hormones is partly determined by changes in the degree of filling of the heart. Since CEBs affect cardiac dynamics, any influence on the release of the hormones may be caused by direct as well as indirect effects. Available evidence thus far suggests little effect of CEBs on the release of antidiuretic hormone or atrial natriuretic factors [Sluiter et al., 1985; Takata et al., 1987]. On the other hand, there are indications that CEBs attenuate the vascular effects of antidiuretic hormone [Hof, 1985, 1986]. This antagonism is of noncompetitive nature, since CEBs do not compete with antidiuretic hormone for the same receptor. It is not yet evident whether this antagonism only occurs at supraphysiological plasma levels of antidiuretic hormone or plays a role in the normal range of ADH concentrations.

Renal Control of the Circulation

The kidney contributes in different ways to the control of the circulation. First, it is a source of some important endocrine systems with an effect on the circulation, in particular the renin-angiotensin system. Second, it is the source of afferent nerves controlling the circulation. The influence of CEBs on these nerves is still unknown. The third way in which the kidney contributes to cardiovascular control is by its influence on body fluid homeostasis. In fact, this is one of the most important long-term mechanisms of blood pressure control.

The effects of CEBs on renal control of body fluid volumes has been the subject of recent reviews [Loutzenhiser and Epstein, 1985; Bauer and Reams, 1987]. We shall therefore only briefly discuss some recent findings of relevance for the hemodynamic effects of CEBs. Evidence discussed above indicates that CEBs have no preferential effect on renal vascular tone. Rather the reduction in renal vascular resistance seems to be determined to a great extent by the autoregulatory adaptation to a reduced perfusion pressure. Thus, renal blood flow and glomerular filtration rate are maintained, as has been observed during acute and chronic treatment of hypertensive patients with verapamil [Leonetti et al., 1980, 1982], diltiazem [Kinoshita et al., 1978; Sunderrajan et al., 1986], nifedipine [Olivari et al., 1979; Leonetti et al., 1982; Guazzi et al., 1984; Bruun et al., 1986] and nitrendipine [Simon and Snyder, 1984; Wallia et al., 1985; Pedrinelli et al.,

1986]. Recent studies indicate that the newer dihydropyridines nicardipine [Van Schayk et al., 1984; Chaignon et al., 1986; Littler et al., 1986] and felodipine [Leonetti et al., 1986] may cause a significant rise in renal blood flow with little effects on glomerular filtration rate in hypertensive patients.

It is not yet clear whether the renal vasodilator action of nicardipine and felodipine is related to additional properties of these drugs or reflect differences in pharmacokinetic behavior, e.g., preferential uptake or binding in the kidney.

A number of studies in animal models show an increase in renal blood flow and glomerular filtration rate following the administration of CEBs [Loutzenhiser and Epstein, 1985; Nagao et al., 1985; Huelsemann et al., 1985; Johns, 1985; Huang, 1986; Bauer and Reams, 1986]. However, one has to consider that the majority of these studies are based upon the use of an anesthetic, which inhibits the baroreflex-mediated sympathetic vasoconstriction in the kidney. In addition, several of these studies were performed in models in which the hypertension depends on renal vasoconstriction by angiotensin II [Huelsemann et al., 1985; Huang, 1986]. In conscious, non-angiotensin-dependent animal models CEBs do not increase renal blood flow or glomerular filtration rate [Nievelstein et al., 1985; Bauer and Reams, 1986]. Felodipine again is an exception, since this agent increases renal blood flow and glomerular filtration rate in the conscious SHR [Nordlander et al., 1985].

Direct infusion of different CEBs in subhypotensive doses into the renal artery of hypertensive patients or experimental animals leads to renal vasodilatation and inhibition of vasoconstrictor responses to angiotensin II [Marre et al., 1982; Dietz et al., 1983; Abe et al., 1983; Loutzenhiser et al., 1985; Ogawa and Ono, 1986; Imagawa et al., 1986a, b; Navar et al., 1986; Steele and Challoner-Hue, 1987]. Blackshear et al. [1987] have reported recently that the renal vasodilator responses to intrarenal infusion are much more pronounced in first-degree relatives of essential hypertensive patients than in control subjects. Since this effect was specific for diltiazem and did not occur in response to acetylcholine, Blackshear et al. [1987] propose the existence of an abnormality in the calcium-dependent control of renal vascular tone in the offspring of essential hypertensive subjects. Since renal vascular tone in these subjects may depend upon the smooth muscle action of angiotensin II [Blackshear et al., 1987], diltiazem exerts its increased renal vasodilator effect probably by inhibiting the calcium-dependent actions of angiotensin II [Loutzenhiser and Epstein,

1985; Bauer and Reams, 1987]. Alternatively, CEBs may have a pronounced effect on renal vasoconstriction induced by the sympathetic nervous system [Nievelstein et al., 1985; Navar et al., 1986; Imagawa et al., 1986b; Pelayo, 1987; Bauer and Reams, 1987; Loutzenhiser et al., 1985, 1987].

In spite of their relative lack of effect on renal blood flow and (GFR) glomerular filtration rate in intact, nonangiotensin-II- or noradrenaline-stimulated individuals, CEBs induce natriuresis and diuresis. Usually, this effect is short-lasting and occurs both in hypertensive humans and experimental animals after the administration of verapamil [Dietz et al., 1983; Bell and Lindner, 1984], diltiazem [Kinoshita et al., 1978; Narita et al., 1983; Johns, 1985; Nagao et al., 1985; Vemulapalli et al., 1987], nifedipine [Leonetti et al., 1982; Yokoyama and Kaburagi, 1983; Johns, 1985; Nagao et al., 1985], nitrendipine [Wallia et al., 1985; Luft et al., 1985; Vemulapalli et al., 1987], nicardipine [Van Schayk et al., 1984; Elliot et al., 1985; Young et al., 1985; Littler et al., 1986; Chaignon et al., 1986], or felodipine [Nordlander et al., 1985; Leonetti et al., 1986]. In the long-term, the diuretic and natriuretic effects disappear, but CEB therapy is usually not associated with generalized fluid and electrolyte retention [cf., Bauer and Reams, 1987]. In some instances, localized ankle edema is reported. However, this effect is probably due to raised capillary pressure as a result of the action of nifedipine on the major limb arteries.

The early diuretic and natriuretic effects of CEBs independent of their renal vascular action point in the direction of direct effects on tubular fluid and electrolyte reabsorption. Such a direct effect of CEBs has indeed been reported in different animal models [Loutzenhiser and Epstein, 1985; Bauer and Reams, 1987]. Recent clearance experiments in essential hypertensive subjects confirm an effect of nifedipine on tubular reabsorption processes [Bruun et al., 1986; Krusell et al., 1987]. An alternative explanation for a direct tubular effect may be that CEBs influence the release of hormones with an effect on tubular fluid and electrolyte reabsorption. There is indeed some evidence for an influence on aldosterone release (see section entitled 'Endocrine Control of the Circulation', p. 44). Another candidate could be the atrial natriuretic factor. The early hemodynamic effects of CEBs certainly favor the release of atrial natriuretic factor from the heart. However, there is as yet no experimental evidence for a role of this factor in the renal actions of CEBs.

In conclusion, a therapeutically very relevant feature of CEB activity is that these drugs in general do not cause retention of fluid and electrolytes as opposed to many classical vasodilators. This effect seems to be independent of vascular effects in intact, nonstimulated (angiotensin II, noradrenaline) individuals. It appears, therefore, that CEBs have prominent effects on tubular reabsorption of fluids and electrolytes, either by a direct action or indirectly via hormones that affect tubular function. The renal effects of CEBs deserve further fundamental research. In addition, studies on the effects of CEBs on the distribution of body fluid volumes could enhance insight in some important aspects of their long-term hemodynamic actions.

Long-Term Control by Structural Design of the Cardiovascular System

Changes in hemodynamic variables in cardiovascular diseases, such as essential hypertension or heart failure, may ultimately be maintained by structural alterations in the heart and vascular system [Lever, 1986]. Although the present goal of antihypertensive therapy or drug treatment of heart failure is improvement of the hemodynamic status, a future goal may well be the reversal of changes in the structural design of the cardiovascular system. Very few attempts have been made thus far to approach the design of new or experimental analysis of existing drugs along those lines.

Several CEBs reduce cardiac hypertrophy due to chronic hypertension [Kobrin et al., 1984; Ruskoaho and Savolainen, 1985; Stasch et al., 1986; Friberg and Nordlander, 1986; Friberg et al., 1986]. This effect can be dissociated from the hemodynamic changes, although the precise molecular mechanism is still unknown [Kobrin et al., 1984].

The effects of CEBs on long-term changes in vascular design are less clear. Nyborg and Mulvany [1985] showed no changes in mesenteric resistance vessel morphology during long-term treatment of SHR with felodipine. On the other hand, Lundin and Hallbäck-Nordlander [1984] using a hemodynamic technique for measuring changes in resistance vessel design, concluded that felodipine can reverse resistance vessel hypertrophy in the SHR. The conflicting results may relate not only to differences in techniques to measure vessel design, but also to the use of SHR of different ages, different vascular beds or the simultaneous administration of other drugs.

CEBs have been attributed antiatherogenic properties [cf., recent review by Weinstein et al., 1987]. High doses of CEBs can significantly decrease atherogenic lesions in cholesterol-fed rabbits [Fleckenstein, 1983; Weinstein et al., 1987]. The mechanisms underlying the vascular 'protective' effect of CEBs are still unknown. Recent studies suggest that nifedipine inhibits arterial smooth muscle cell proliferation [Nilsson et al.,

1985]. Furthermore, verapamil and diltiazem inhibit platelet activation [Addonizio et al., 1986]. However, it is not yet clear whether these actions are specific for the investigated drugs or are general properties of CEBs. In view of the high prevalence of coronary heart diseases, the antiatherogenic properties of CEBs deserve further attention.

Synthesis: The Overall Hemodynamic Effects of Calcium Entry Blockers

The preceding paragraphs described the effects of CEBs on the basic elements of the cardiovascular system – the heart and blood vessels – as well as on the major cardiovascular control mechanisms. It is now possible to synthesize this information into an overall picture of the hemodynamic effects of CEBs. Figure 2 attempts to give such a picture for the hemodynamic effects of CEBs in essential hypertension. Two patterns of activity are given: (a) CEBs with a predominant vascular effect, in particular the dihydropyridines, and (b) CEBs with additional cardiac effects, in particular verapamil and diltiazem.

Important summarizing features of the hemodynamic action of CEBs in hypertension are:

- The early fall in blood pressure is caused by a reduction of peripheral vascular resistance.
- The fall in blood pressure and vascular resistance trigger a sino-aortic baroreflex-mediated rise in sympathetic activity. With dihydropyridines this leads to an early rise in HR and CO. The direct cardiac actions of verapamil and diltiazem, in addition to their possible interference with the baroreflex, prevent a rise in HR and CO. In the long-term, the baroreflex adapts to the prevailing pressure level, thus allowing HR and CO to return to values close to (dihydropyridines) or just below (verapamil and diltiazem) control levels.
- The peripheral vasodilator effect of CEBs is not homogeneously distributed over all vascular beds. Vasodilatation occurs especially in the skeletal muscle and coronary vascular beds. The gastrointestinal, cerebral and renal beds are also dilated, whereas the skin is not affected. Vascular inhomogeneity does not seem due to differences in vascular CEB 'receptor' mechanisms, but can be regarded as the consequence of differences in the control of vascular resistances. CEBs act in two fundamentally different modes on vascular resistance, depending upon

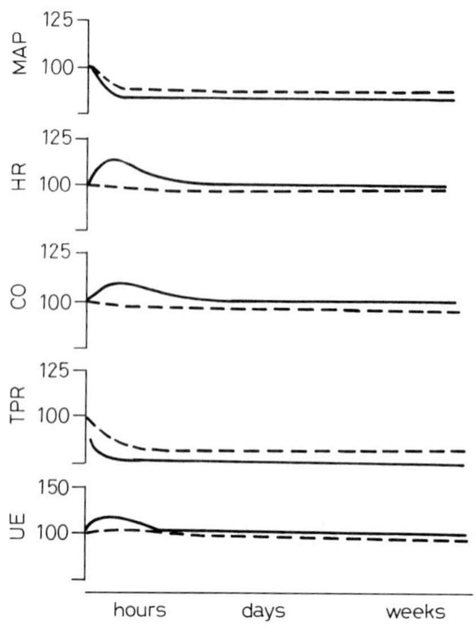

Fig. 2. Dynamics of the effects of different CEBs. Changes are expressed as percentages of pretreatment values. Solid lines: dihydropyridines; interrupted lines: verapamil and diltiazem. MAP=Mean arterial pressure; HR=heart rate; CO=cardiac output; TPR=total peripheral resistance; UE=urine excretion.

whether or not they are engaging the sympathetic nervous system via the baroreceptor reflex.

- The long-term antihypertensive action of CEBs is not counteracted by renal fluid retention, as occurs with classical vasodilators. The lack of renal fluid retention is not due to a renal vascular effect of CEBs, but seems related to a direct tubular action or to the influence of CEBs on hormones influencing tubular transport mechanisms.

This hemodynamic profile of activity has given the CEBs a prominent place in the treatment of hypertension. The cardiac as well as afterload effects of CEBs have contributed to their long-standing position in the treatment of angina pectoris. More recently, the use of CEBs in the treatment of chronic congestive heart failure has been a matter of debate [cf., Drexler and Just, 1986]. Vasodilators can produce a useful improvement in the hemodynamics of heart failure, as a consequence of their afterload reducing effects. However, the cardiac effects of CEBs – especially verapamil and diltiazem – may not favor their use in congestive heart failure. On the other hand, a clear advantage of CEBs over classical vasodilators is their relative lack of renal fluid retention. On-going clinical trials on the use of CEBs should define their potential in the treatment of heart failure.

A final cardiovascular disease with potential therapeutic application of CEBs is Raynaud's phenomenon. The analysis presented above does not suggest use of CEBs in the treatment of this disorder, since CEBs have little or no effect on skin circulation. Results of clinical trials on the use of CEBs in the treatment of Raynaud's disease have indeed led to rather variable results [cf., Wigley et al., 1987].

References

Abe, Y.; Komori, T.; Miura, K.; Takada, T.; Imanishi, M.; Okahara, T.; Yamamoto, K.: Effects of the calcium antagonist nicardipine on renal function and renin release in dogs. J. cardiovasc. Pharmacol. *5:* 254–259 (1983).

Addonizio, V.P.; Fisher, C.A.; Strauss, J.F.; Wachtfogel, Y.T.; Colman, R.W.; Josephson, M.E.: Effects of verapamil and diltiazem on human platelet function. Am. J. Physiol. *250:* H366–H371 (1986).

Agner, E.; Rehling, M.; Trap-Jensen, J.: Hemodynamic effects of single-dose felodipine in normal man. Drugs *29:* suppl. 2, pp. 36–40 (1985).

Alabaster, V.A.; Solca, A.M.: Lack of differential inhibition by nifedipine of pressor responses induced by alpha-1- and alpha-2-adrenoceptor agonists and by angiotensin II in anesthetized cats. Clin Sci. mol. Med. *68:* suppl. 10, pp. 73–75 (1985).

Altura, B.M.; Altura, B.T.; Gebrewold, A.: Differential effects of the calcium antagonist verapamil on lumen sizes of terminal arterioles and muscular venules in the rat mesenteric, pial and skeletal muscle microvasculatures. Br. J. Pharmacol. *70:* 351–353 (1980).

Amodeo, C.; Kobrin, I.; Ventura, H.O.; Messerli, F.H.; Frohlich, E.D.: Immediate and short-term hemodynamic effects of diltiazem in patients with hypertension. Circulation *73:* 108–113 (1986).

Anderson, O.; Granērus, G.; Hedner, T.: Felodipine: a calcium inhibiting vasodilator in refractory hypertension. Drugs *29:* suppl. 2, pp. 102–108 (1985).

Aoki, K.; Asano, M.: Effects of Bay K 8644 and nifedipine on femoral arteries of spontaneously hypertensive rats. Br. J. Pharmacol. *88:* 221–230 (1986).

Aoki, K.; Sato, K.; Kondo, S.; Yamamoto, M.: Hypotensive effects of diltiazem to normals and essential hypertensives. Eur. J. clin. Pharmacol. *25:* 475–480 (1983).

Armstead, W.M.; Lippton, H.L.; Hyman, A.L.; Kadowitz, P.J.: Influence of nisoldipine on vascular resistance and vasoconstrictor responses in cats. Am. J. Physiol. *252:* H816–H825 (1987).

Bache, R.J.; Tockman, B.A.: Effect of nitroglycerin and nifedipine on subendocardial perfusion in the presence of a flow-limiting coronary stenosis in the awake dog. Circulation Res. *50:* 678–687 (1982).

Bauer, J.H.; Reams, G.: Short- and long-term effects of calcium entry blockers on the kidney. Am. J. Cardiol. *59:* 66A–71A (1987).

Bell, A.J.; Lindner, A.: Effects of verapamil and nifedipine on renal function and hemodynamics in the dog. Renal Physiol. *7:* 329–343 (1984).

Bevan, J.A.: Selective action of diltiazem on cerebral vascular smooth muscle in the rabbit: antagonism of extrinsic but not intrinsic maintained tone. Am. J. Cardiol. *49:* 519–524 (1982).

Bevan, J.A.; Bevan, R.D.; Hwa, J.J.; Owen, M.P.; Tayo, F.M.: Calcium regulation in vascular smooth muscle: is there a pattern to its variability within the arterial tree? J. cardiovasc. Pharmacol. *8:* suppl. 8, pp. 71–75 (1986).

Bijak, A.; Pasternac, A.; McPherson, G.A.; Bevan, J.A.: Vascular bed and vasoconstrictor-dependent selectivity of the calcium channel antagonist, PN 200–110. Eur. J. Pharmacol. *132:* 313–317 (1986).

Blackshear, J.L.; Garnic, D.; Williams, G.H.; Harrington, D.P.; Hollenberg, N.K.: Exaggerated renal vasodilator response to calcium entry blockade in first-degree relatives of essential hypertensive subjects. Hypertension *9:* 384–389 (1987).

Bohr, D.F.: Vascular smooth muscle: dual effect of calcium. Science *139:* 597–599 (1963).

Bolt, G.R.; Saxena, P.R.: Acute systemic and regional hemodynamic effects of felodipine, a new calcium antagonist, in conscious renal hypertensive rabbits. J. cardiovasc. Pharmacol. *6:* 707–712 (1984).

Bolton, T.B.: Mechanisms of action of transmitters and other substances on smooth muscle. Physiol. Rev. *3:* 606–718 (1979).

Bonzel, T.; Wollschläger, H.; Löllgen, H.; Münzel, T.; Behrle, B.; Just, H.: Effect of intravenous diltiazem on coronary artery diameter in patients with coronary artery disease. J. Am. Coll. Cardiol. *3:* suppl., 578 (1984).

Bou, J.; Massingham, R.: Effect of diltiazem upon contractile response to phenylephrine, cirazoline, Sgd 101/75, St 587 and B-HT 920 in rabbit aorta and dog saphenous vein preparations. Eur. J. Pharmacol. *121:* 319–325 (1986).

Brockaert, A.; Godfraind, T.: A comparison of the inhibitory effect of cinnarazine and papaverine on the noradrenaline and calcium-evoked contraction of isolated rabbit aorta and mesenteric arteries. Eur. J. Pharmacol. *53:* 281–288 (1979).

Brown, B.G.; Bolson, E.; Petersen, R.B.; Pierce, C.D.; Dodge, H.T.: The mechanisms of nitroglycerin action: stenosis vasodilation as a major component of the drug response. Circulation *64:* 1089–1097 (1981a).

Brown, B.G.; Pierce, C.D.; Petersen, R.B.; Singh, B.N.; Bolson, E.L.; Dodge, H.T.: Verapamil, a mild epicardial coronary dilator, inhibits sympathetic and ergonovine-induced coronary constriction in humans. Circulation *64:* suppl. IV, p. 150 (1981b).

Bruun, N.E.; Ibsen, H.; Skott, P.; Giese, J.; Holstein-Rathlou, N.H.: Redistribution of renal tubular sodium resorption during long-term treatment with nifedipine in patients with essential hypertension. J. Hypertens. *4:* suppl. 6, pp. 177–179 (1986).

Burnstock, G.: The changing face of autonomic neurotransmission. Acta physiol. scand. *126:* 67–91 (1986).

Campbell, B.C.; Kelman, A.W.; Hillis, W.S.: Non-invasive assessment of the hemodynamic effects of nicardipine in normotensive subjects. Br. J. clin. Pharmacol. *20:* suppl., pp. 55–61 (1985).

Cauvin, C.; Hwang, O.; Yamamoto, M.; Van Breemen, C.: Effects of dihydropyridines on tension and calcium-45 influx in isolated mesenteric resistance vessels from spontaneously hypertensive and normotensive rats. Am. J. Cardiol. *59:* 116B–112B (1987).

Cauvin, C.; Loutzenhiser, R.; Van Breemen, C.: Mechanisms of calcium antagonist induced vasodilation. Annu. Rev. Pharmacol. Ther. *23:* 373–396 (1983).

Chaffmann, M.; Brogden, R.N.: Diltiazem: a review of its pharmacological properties and therapeutic efficacy. Drugs *29:* 387–454 (1985).

Chaignon, M.; Bellet, M.; Lucsko, M.; Rapoud, C.; Guedon, J.: Acute and chronic effects of a new calcium inhibitor, nicardipine, on renal hemodynamics in hypertension. J. cardiovasc. Pharmacol. *8:* 892–897 (1986).

Charlton, J.D.; Baertschi, A.J.: Responses of aortic baroreceptors to changes of aortic blood flow and pressure in rat. Am. J. Physiol. *242:* H520–H524 (1982).

Chiu, J.H.: Differential sensitivity of calcium channels to dihydropyridines. The modulated receptor hypothesis. Biochem. Pharmac. *35:* 4115–4120 (1986).

Chiu, A.T.; McCall, D.E.; Thoolen, M.J.M.C.; Timmermans, P.B.M.W.M.: Ca^{++} utilization in the constriction of rat aorta to full and partial $alpha_1$-adrenoceptor agonists. J. Pharmac. exp. Ther. *238:* 224–231 (1986).

Clarke, G.; Davies, G.J.; Kerwin, R.; Hackett, D.; Larkin, S.; Dawbarn, D.; Lee, Y.; Bloom, S.R.; Yacoub, M.; Maseri, A.: Coronary artery infusion of neuropeptide Y in patients with angina pectoris. Lancet *i:* 1057–1059 (1987).

Cody, R.J.: The hemodynamics of calcium channel antagonists in hypertension: vascular and myocardial responses. Circulation *75:* 1175–1179 (1987).

Colantuoni, A.; Bertuglia, S.; Intaglietta, M.: The effects of alpha- or beta-adrenergic receptor agonists and antagonists and calcium entry blockers on the spontaneous vasomotion. Microvasc. Res. *28:* 143–158 (1984).

Cooke, J.P.; Rimele, T.J.; Flavahan, N.A.; Vanhoutte, P.M.: Nimodipine and inhibition of alpha-adrenergic activation of the isolated canine saphenous vein. J. Pharmac. exp. Ther. *234:* 598–602 (1985).

Coruzzi, P.; Biggi, A.; Musiari, L.; Ravanetti, C.; Novarini, A.: Cardiovascular, baroreflex and humoral responses in hypertensive patients during nicardipine therapy. Eur. J. clin. Pharmacol. *29:* 371–374 (1985).

De Langen, C.D.; Meijboom, E.J.; Kingma, J.H.: Heart rate dependent atrioventricular nodal conduction and the effects of calcium channel blocking drugs: comparison of verapamil and nifedipine. J. cardiovasc. Pharmacol. *6:* 267–272 (1984).

De Mey, J.G.; Vanhoutte, P.M.: Uneven distribution of postjunctional $alpha_1$- and $alpha_2$-like adrenoceptors in canine arterial and venous smooth muscle. Circulation Res. *48:* 875–884 (1981).

Dietz, J.R.; Davis, J.O.; Freeman, R.H.; Villareal, D.; Echtenkamp, S.F.: Effects of intrarenal infusion of calcium entry blockers in anesthetized dogs. Hypertension *5:* 482–488 (1983).

Drexler, H.; Flaim, S.F.; Fields, R.H.; Zelis, R.: Effects of nisoldipine on cardiocirculatory dynamics and cardiac output distribution in conscious rats at rest and during treadmill exercise. J. Pharmac. exp. Ther. *232:* 376–381 (1985).

Drexler, H.; Just, H.: Kalziumantagonisten bei Herzinsuffizienz? Klin. Wschr. *64:* 993–1002 (1986).

Drexler, H.; Truog, A.G.; Zelis, R.; Flaim, S.F.: Regional vascular and hemodynamic effects of orally administered nisoldipine in conscious rats. J. cardiovasc. Pharmacol. *8:* 151–155 (1986).

Duncker, D.J.; Hartog, J.M.; Hugenholtz, P.G.; Saxena, P.R.; Verdouw, P.D.: The effects of nisoldipine (Bay K 5552) on cardiovascular performance and regional blood flow in pentobarbital-anesthetized pigs with or without beta-adrenoceptor blockade. Br. J. Pharmacol. *88:* 9–18 (1986a).

Duncker, D.J.; Heiligers, J.; Mylecharane, E.J.; Saxena, P.R.; Verdouw, P.D.: Nimodipine-induced changes in the distribution of carotid blood flow and cardiac output in pentobarbitone-anesthetized pigs. Br. J. Pharmacol. *89:* 35–46 (1986b).

Elliott, H.F.; Pasani, F.; Reid, J.L.: Effects of nicardipine on aldosterone release and pressor mechanisms. Br. J. clin. Pharmacol. *20:* 995–1025 (1985).

Fagard, R.; Lijnen, P.; Moerman, E.; Staessen, J.; Amery, A.: Acute hemodynamic and humoral responses to felodipine and metoprolol in mild hypertension. Eur. J. clin. Pharmacol. *32:* 71–75 (1987).

Fam, W.M.; McGregor, M.: Effect of nitroglycerin and dipyridamole on regional coronary resistance. Circulation Res. *22:* 649–655 (1968).

Ferguson, D.W.; Dorsey, J.K.: Effects of nifedipine on baroreflex modulation of vascular resistance in man. Am. Heart J. *109:* 55–62 (1985).

Finch, M.B.; Johnston, G.D.: The acute peripheral vascular effects of diltiazem in man: dose response characteristics. Br. J. Pharmacol. *19:* 131P (1985).

Flaim, S.F.; Newman, E.D.; Annibali, J.A.: Effects of intravenous diltiazem on cardiocirculatory dynamics and cardiac output distribution in conscious spontaneously hypertensive rats. J. cardiovasc. Pharmacol. *8:* 241–251 (1986).

Flaim, S.F.; Zelis, R.: Calcium blockers. Mechanisms of action and clinical applications (Urban & Schwarzenberg, Baltimore 1982a).

Flaim, S.F.; Zelis, R.: Effects of diltiazem on total cardiac output distribution in conscious rats. J. Pharmac. exp. Ther. *222:* 359–366 (1982b).

Fleckenstein, A.: Specific pharmacology of calcium in myocardium, cardiac pacemakers, and vascular smooth muscle. Annu. Rev. Pharmacol. Toxicol. *17:* 149–166 (1977).

Fleckenstein, A.: Calcium antagonism in heart and smooth muscle, pp. 1–398 (Wiley, New York 1983).

Fouad, F.M.; Pedrinelli, R.; Bravo, E.L.; Abi-Samra, F.; Textor, S.C.; Tarazi, R.C.: Clinical and systemic hemodynamic effects of nitrendipine. Clin. Pharmacol. Ther. *35:* 768–775 (1984).

Friberg, P.; Folkow, B.; Nordlander, M.: Cardiac dimensions in spontaneously hypertensive rats following different modes of blood pressure reduction by antihypertensive treatment. J. Hypertens. *4:* 85–92 (1986).

Friberg, P.; Nordlander, M.: Influence of long-term antihypertensive therapy on cardiac function, coronary flow and myocardial oxygen consumption in spontaneously hypertensive rats. J. Hypertens. *4:* 165–173 (1986).

Frohlich, E.D.: Hemodynamic effects of calcium entry blocking agents in normal and hypertensive rats and man. Am. J. Cardiol. *56:* 21H–27H (1985).

Fujimoto, S.; Oaki, K.; Matsuda, T.: Comparative effects of bepridil and diltiazem on systemic blood pressure and isolated peripheral arteries in spontaneously hypertensive rats. Eur. J. Pharmacol. *124:* 243–254 (1986).

Furchgott, R.F.; Zazwadzki, J.V.: The obligatory role of endothelial cells in the relaxation of arterial smooth muscle by acetylcholine. Nature *288:* 373–376 (1980).

Gerold, M.; Haeusler, G.: Alpha-2-adrenoceptors in rat resistance vessels. Arch. Pharmacol. *322:* 29–33 (1983).

Goa, K.L.; Sorkin, E.M.: Nitrendipine: a review of its pharmacodynamic and pharmacokinetic properties, and therapeutic efficacy in the treatment of hypertension. Drugs *33:* 123–155 (1987).

Godfraind, T.: Mechanisms of action of calcium entry blockers. Fed. Proc. *40:* 2866–2871 (1981).

Godfraind, T.: Actions of nifedipine on calcium fluxes and contraction in isolated rat arteries. J. Pharmac. exp. Ther. *224:* 443–450 (1983).

Godfraind, T.; Miller, R.C.: Actions of prostaglandin F_2 and noradrenaline on calcium exchange and contraction in rat mesenteric arteries and their sensitivity to calcium entry blockers. Br. J. Pharmacol. *72:* 229–236 (1982).

Godfraind, T.; Miller, R.C.; Socrates Lima, J.: Selective alpha-1- and alpha-2-adrenoceptor agonist induced contractions and ^{45}Ca fluxes in the rat isolated aorta. Br. J. Pharmacol. *77:* 597–604 (1982).

Gross, R.; Kirchheim, H.; Von Olshausen, K.: Effects of nifedipine on coronary and systemic hemodynamics in the conscious dog. Arzneimittel-Forsch. *29:* 1361–1368 (1979).

Guazzi, M.D.; De Cesare, N.; Galli, C.; Salvioni, A.; Tramontana, C.; Tamborini, G.; Bartorelli, C.: Calcium channel blockade with nifedipine and angiotensin converting enzyme inhibition with captopril in the therapy of patients with severe primary hypertension. Circulation *70:* 279–284 (1984).

Guthrie, G.P.; McAllister, R.G.; Kotchen, T.A.: Effects of intravenous and oral verapamil upon pressure and adrenal steroidogenic responses in normal man. J. clin. Endocr. Metab. *57:* 339–343 (1983).

Guyton, A.C.: Textbook of medical physiology (Saunders, Philadelphia 1986).

Hamilton, C.A.; Reid, J.L.; Summer, D.J.: Acute effects of phenoxybenzamine on alpha-adrenoceptor response in vivo and in vitro: relation of in-vivo pressor responses to the number of specific adrenoceptor binding sites. J. Cardiovasc. Pharmacol. *5:* 868–873 (1983).

Haws, C.W.; Heistad, D.D.: Effects of nimodipine on cerebral vasoconstrictor responses. Am. J. Physiol. *247:* H170–H176 (1984).

Hayashi, M.; Aizawa, Y.; Satoh, M.; Suzuki, K.; Shibata, A.: Effect of nifedipine on neuropeptide Y induced coronary vasoconstriction in anesthetized dogs. Jap. Heart J. *27:* 251–257 (1986).

Heesch, C.M.; Miller, B.M.; Thames, M.D.; Abboud, F.M.: Effects of calcium channel blockers on isolated carotid baroreceptors and baroreflex. Am. J. Physiol. *245:* H653–H658 (1983).

Hicks, P.E.; Tierney, C.; Langer, S.Z.: Preferential antagonism by diltiazem of alpha-2-adrenoceptor mediated vasoconstrictor responses in perfused tail arteries of spontaneously hypertensive rats. Arch. Pharmacol. *328:* 388–395 (1985).

Higuchi, S.; Takeshita, A.; Ito, N.; Imaizumi, T.; Matsuguchi, H.; Nakamura, M.: Arterial pressure and heart rate responses to calcium channel blockers administered in the brainstem in rats. Circulation Res. *57:* 244–251 (1985).

Hiley, C.R.; Thomas, G.R.: Effects of alpha-adrenoceptors agonists on cardiac output and its regional distribution in the pithed rat. Br. J. Pharmacol. *90:* 61–70 (1987).

Hiramatsu, K.; Yamagishi, F.; Kubota, T.; Yamada, T.: Acute effects of the calcium antagonist nifedipine on blood pressure, pulse rate, and the renin-angiotensin-aldosterone system in patients with essential hypertension. Am. Heart J. *104:* 1346–1350 (1982).

Hof, R.P.: Calcium antagonists and the peripheral circulation: differences and similarities between PY 108–068, nicardipine, verapamil and diltiazem. Br. J. Pharmacol. *78:* 375–394 (1983).

Hof, R. P.: The calcium antagonists PY 108–068 and verapamil diminish the effects of angiotensin II: sites of interaction in the peripheral circulation of anesthetized cats. Br. J. Pharmacol. *82:* 51–60 (1984a).

Hof, R. P.: Selective effects of different calcium antagonists on the peripheral circulation. Trends pharmacol. Sci. *5:* 100–102 (1984b).

Hof, R. P.: Modification of vasopressin- and angiotensin-II-induced changes by calcium antagonists in the peripheral circulation of anesthetized rabbits. Br. J. Pharmacol. *85:* 75–87 (1985).

Hof, R. P.: Vasopressin-induced myocardial depression is neurally mediated and not due to impaired coronary blood flow. Br. J. Pharmacol. *87:* 611–618 (1986).

Hof, R. P.; Evenou, J. P.; Hof-Miyashita, A.: Similar increase in circulating renin after equihypotensive doses of nitroprusside, dihydralazine, or isradipine in conscious rabbits. Eur. J. Pharmacol. *136:* 251–254 (1987).

Hof, R. P.; Hof, A.; Ruegg, U. T.; Cook, N. S.; Vogel, A.: Stereoselectivity at the calcium channel: different profiles of hemodynamic activity of the enantiomers of the dihydropyridine derivative PN 200–110. J. cardiovasc. Pharmacol. *8:* 221–226 (1986).

Hof, R. P.; Scholtysik, G.; Loutzenhiser, R.; Vuorela, H. J.; Neumann, P.: PN 200–110, a new calcium antagonist: electrophysiological inotropic, and chronotropic effects on guinea pig myocardial tissue and effects on contraction and calcium uptake of rabbit aorta. J. cardiovasc. Pharmacol. *6:* 399–406 (1984a).

Hof, R. P.; Scholtysik, G.; Menninger, K.: Effects of the new calcium antagonist PN 200–110 on the myocardium and the regional peripheral circulation in anesthetized cats and dogs. J. cardiovasc. Pharmacol. *6:* 407–416 (1984b).

Hof, R. P.; Vuorela, H. J.; Neumann, P.: PY 108–068, a new potent and selective inhibitor of calcium-induced contraction of rabbit aortic rings. J. cardiovasc. Pharmacol. *4:* 344–351 (1982).

Holck, M.; Gerold, M.: Calcium entry blockers inhibit vasoconstrictor responses to sympathetic nerve stimulation mediated by alpha-1-adrenoceptors. Arch. Pharmacol. *330:* 22–32 (1985).

Hossack, K. F.; Brown, B. G.; Stewart, D. K.; Dodge, H. T.: Diltiazem-induced blockade of sympathetically mediated constriction of normal and diseased coronary arteries. Circulation *70:* 465–471 (1984).

Huang, W. C.: Effects of verapamil alone and with captopril on blood pressure and bilateral renal function in Goldblatt hypertensive rats. Clin. Sci. mol. Med. *70:* 453–460 (1986).

Huelsemann, J. L.; Sterzel, R. B.; McKenzie, D. E.; Wilcox, C. S.: Effects of a calcium entry blocker on blood pressure and renal function during angiotensin-induced hypertension. Hypertension *7:* 374–379 (1985).

Hurwitz, L.: Pharmacology of calcium channels and smooth muscle. Annu. Rev. Pharmacol. Toxicol. *26:* 225–258 (1986).

Hwa, J. J.; Bevan, J. A.: A nimodipine-resistant calcium pathway is involved in myogenic tone in a resistance artery. Am. J. Physiol. *251:* H182–H189 (1986).

Imagawa, J.; Kurosawa, H.; Satoh, S.: Effects of nifedipine on renin release and renal function in anesthetized dogs. J. cardiovasc. Pharmacol. *8:* 636–640 (1986a).

Imagawa, J.; Kusaba-Suzuki, M.; Satoh, S.: Preferential inhibitory effect of nifedipine on angiotensin-II-induced renal vasoconstriction. Hypertension *8:* 897–903 (1986b).

Iyoda, I.; Takahashi, H.; Takeda, K.; Inoue, A.; Yoneda, S.; Sasaki, S.; Okajima, H.; Yoshimura, M.; Ijichi, H.: Centrally induced vasodepressor responses to diltiazem, a calcium channel blocker, in rats. J. Hypertens. *3:* 639–644 (1985).

Jandhyala, B. S.: Studies on the effects of felodipine on the vascular resistance and reactivity in pentobarbital anesthetized dogs. Arch. Pharmacol. *329:* 398–403 (1985).

Jayakody, R. L.; Kappagoda, C. T.; Senaratne, M. P. J.; Sreeharan, N.: Absence of effect of calcium antagonists on endothelium-dependent relaxation in rabbit aorta. Br. J. Pharmacol. *91:* 155–164 (1987a).

Jayakody, R. L.; Senaratne, M. P. J.; Kappagoda, C. T.: Effects of nicardipine and diltiazem on alpha-adrenoceptor responses in canine saphenous veins. Eur. J. Pharmacol. *136:* 269–277 (1987b).

Jie, K.; Van Brummelen, P.; Timmermans, P. B. M. W.; Thoolen, M. J. M. C.; Van Zwieten, P. A.: Effects of PN 200–110, a new calcium antagonist, on alpha-1- and alpha-2-adrenoceptor mediated vasoconstriction elicited by selective alpha-adrenoceptor agonists and catecholiamines in the pithed rat. Archs int. Pharmacodyn. Ther. *278:* 72–86 (1985).

Jim, K. F.; DeMaranis, R. M.; Matthews, W. D.: Measurements of $^{45}Ca^{2+}$ uptake and contractile responses after activation of postsynaptic alpha-1-adrenoceptors in the isolated canine saphenous vein: effects of calcium entry blockade. Eur. J. Pharmacol. *107:* 199–208 (1985).

Jim, K. F.; Macia R. A.; Matthews, W. D.: Role of receptor reserve in the inhibition of alpha-1-adrenoceptor-mediated pressor responses by calcium antagonists in the pithed rat. J. Pharmac. exp. Ther. *238:* 89–94 (1986).

Johns, E. J.: The influence of diltiazem and nifedipine on renal function in the rat. Br. J. Pharmacol. *84:* 707–713 (1985).

Joyal, M.; Cremer, K. F.; Pieper, J. A.; Feldman, R. L.; Pepine, C. J.: Systemic, left ventricular and coronary hemodynamic effects of intravenous diltiazem in coronary artery disease. Am. J. Cardiol. *56:* 413–417 (1985).

Julou, G.; Freslon, J. L.: Effects of calcium entry blockers on Ca^{2+}-induced contraction of depolarized and noradrenaline-exposed rat resistance vessels. Eur. J. Pharmacol. *129:* 261–270 (1986).

Kanda, K.; Flaim, S. F.: Effects of nifedipine on total cardiac output distribution in conscious rat. J. Pharmac. exp. Ther. *228:* 711–718 (1984).

Kannan, M. S.; Seip, A. E.: Mechanisms of alpha-adrenoceptor-mediated contractions of rat mesenteric artery. Can. J. Physiol. Pharmacol. *64:* 419–423 (1986).

Kannan, M. S.; Seip, A. E.; Crankshaw, D. J.: Effect of nifedipine on calcium-induced contractions to potassium in the aorta and mesenteric arteries of spontaneously hypertensive and normotensive rats. Can. J. Physiol. Pharmacol. *64:* 310–314 (1986).

Kazda, S.; Garthoff, B.; Meyer, H.; Schlossmann, K.; Stoepel, K.; Towart, R.; Vater, W.; Wehinger, E.: Pharmacology of a new calcium antagonistic compound, isobutyl methyl 1,4-dihydro-2,6-dimethyl-4-(2-nitrophenyl)-3,5-pyridinedicarboxylate (nisoldipine, Bay k5552). Arzneimittel-Forsch. *30:* 2144–2162 (1980).

Kazda, S.; Garthoff, B.; Knorr, A.: Nitrendipine and other calcium entry blockers (calcium antagonists) in hypertension. Fed. Proc. *42:* 196–200 (1983).

Kinoshita, M.; Kusukawa, R.; Shimono, Y.; Motomura, M.; Tomonaga, G.; Hoshino, T.: Effects of diltiazem on renal hemodynamics and urinary electrolyte excretion. Jap. Circul. J. *42:* 553–560 (1978).

Kiowski, H.; Bertel, S.; Erne, P.; Bolli, P.; Hulthén, W.L.; Ritz, R.; Bühler, F.R.: Hemodynamic and reflex responses to acute and chronic antihypertensive therapy with the calcium entry blocker nifedipine. Hypertension *5:* suppl. I, pp. 70–76 (1983).

Kleinbloesem, C.H.; Van Brummelen, P.; Danhof, M.; Faber, H.; Urquhart, J.; Breimer, D.D.: Rate of increase in the plasma concentration of nifedipine as a major determinant of its hemodynamic effects in humans. Clin. Pharmacol. Ther. *41:* 26–30 (1987).

Knight, D.R.; Kirby, D.A.; Vatner, S.F.: Effects of a calcium channel blocker on cardiac output distribution in conscious hypertensive dogs. Hypertension *7:* 380–385 (1985).

Kobrin, I.; Sesoko, S.; Pegram, B.L.; Frohlich, E.D.: Reduced cardiac mass by nitrendipine is dissociated from systemic or regional hemodynamic changes in rats. Cardiovasc. Res. *18:* 158–162 (1984).

Kojima, K.; Kojima, I.; Rasmussen, H.: Dihydropyridine calcium agonist and antagonist effects on aldosterone secretion. Am. J. Physiol. *247:* E645–E650 (1984).

Krusell, L.R.; Christensen, C.K.; Pedersen, O.L.: Acute natriuretic effect of nifedipine in hypertensive patients and normotensive controls. A proximal tubular effect. Eur. J. Pharmacol. *32:* 121–126 (1987).

Kubo, K.; Van de Water, A.; Wouters, L.; Xhonneux, R.; Reneman, R.S.: Cardiac and hemodynamic effects of intravenous administrations of flunarizine in closed-chest anesthetized mongrel dogs. Arzneimittel-Forsch. *34:* 434–439 (1984).

Kunze, D.L.; Andresen, M.C.; Torres, L.A.: Do calcium antagonists act directly on calcium channels to alter baroreceptor function? J. Pharmac. exp. Ther. *239:* 303–310 (1986).

Lambert, C.R.; Hill, J.A.; Nichols, W.W.; Feldman, R.L.; Pepine, C.J.: Coronary and systemic hemodynamic effects of nicardipine. Am. J. Cardiol. *55:* 652–656 (1985).

Lappe, R.W.; Barron, K.W.; Faber, J.E.; Brody, M.J.: Selective antagonism of humoral versus neural vasoconstrictor responses by nisoldipine. Hypertension *7:* 216–222 (1985).

Laurent, S.; Girerd, X.; Tsoukaris-Kupfer, D.; Legrand, M.; Huchet-Brisac, A.M.; Schmitt, H.: Opposite central cardiovascular effects of nifedipine and Bay K8644 in anesthetized rats. Hypertension *9:* 132–138 (1987).

Lederballe-Pedersen, O.; Mikkelsen, E.; Christensen, N.J.; Kornerup, H.J.; Pedersen, E.B.: Effect of nifedipine on plasma renin, aldosterone and catecholamines in arterial hypertension. Eur. J. clin. Pharmacol. *15:* 235–240 (1979).

Legrand, V.; Hastir, F.; Vandormael, M.; Collignon, P.; Kulbertus, H.E.: Hemodynamic effects of intravenous diltiazem at rest and exercise in patients with coronary artery disease. Eur. Heart J. *5:* 456–463 (1984).

Lehmann, H.-U.; Hochrein, H.; Witt, E.; Mies, H.W.: Hemodynamic effects of calcium antagonists. Hypertension *5:* suppl. II, pp. 66–73 (1983).

Leonetti, G.; Cuspidi, C.; Sampieri, L.; Terzoli, L.; Zanchetti, A.: Comparison of cardiovascular, renal and humoral effects of acute administration of two calcium channel blockers in normotensive and hypertensive subjects. J. cardiovasc. Pharmacol. *4:* suppl., pp. 319–324 (1982).

Leonetti, G.; Gradnik, R.; Terzoli, L.; Fruscio, M.; Rupoli, L.; Cuspidi, C.; Sampieri, L.; Zanchetti, A.: Effects of single and repeated doses of the calcium antagonist felodipine on blood pressure, renal function, electrolytes and water balance, and reninangiotensin-aldosterone system in hypertensive patients. J. cardiovasc. Pharmacol. *8:* 1243–1248 (1986).

Leonetti, G.; Sala, C.; Bianchini, C.; Terzoli, L.; Zanchetti, A.: Antihypertensive and renal effects of orally administered verapamil. Eur. J. clin. Pharmacol. *18:* 375–382 (1980).

Levenson, J.; Simon, A.C.; Bouthier, J.; Maarek, B.C.; Safar, M.E.: The effect of acute and chronic nicardipine therapy on forearm arterial hemodynamics in essential hypertension. J. clin. Pharmacol. *20:* suppl. 1, pp. 107–113 (1985).

Levenson, J.; Simon, A.C.; Safar, M.E.; Bouthier, J.; Maarek, B.C.: Large arteries in hypertension: acute effects of a new calcium entry blocker, nitrendipine. J. cardiovasc. Pharmacol. *6:* suppl. 7, pp. 1006–1010 (1984).

Lever, A.F.: Slow pressor mechanisms in hypertension: a role for hypertrophy of resistance vessels. J. Hypertens. *4:* 515–524 (1986).

Lew, M.J.; Angus, J.A.: Alpha-1- and alpha-2-adrenoceptor mediated pressor responses: are they differentiated by calcium antagonists or by functional antagonism. J. cardiovasc. Pharmacol. *7:* 401–408 (1985).

Lewis, G.R.J.: Verapamil in the management of chronic hypertension. Clin. invest. Med. *3:* 175–177 (1980).

Littler, W.A.; Young, M.A.; Smith, S.A.: The acute and chronic effects of nicardipine on blood pressure, heart rate, neurocirculatory reflexes and renal function in hypertensive patients. Br. J. clin. Pharmacol. *22:* suppl., pp. 239–241 (1986).

Ljung, B.: Vascular selectivity of felodipine. Drugs *29:* suppl. 2, pp. 46–58 (1985).

Long, C.J.; Stone, T.W.: The release of endothelium-derived relaxant factor is calcium-dependent. Blood Vessels, *22:* 205–208 (1985).

Lopez, L.M.; Mehta, J.L.; Baz, R.; Agulla, E.: Effects of nitrendipine and hydralazine on plasma catecholamines in essential hypertension. Clin. Pharmacol. Ther. *36:* 444–450 (1984).

Loutzenhiser, R.; Epstein, M.: Effects of calcium antagonists on renal hemodynamics. Am. J. Physiol. *249:* F619–F629 (1985).

Loutzenhiser, R.; Epstein, M.; Horton, C.: Inhibition by diltiazem of pressure-induced afferent vasoconstriction in the isolated perfused rat kidney. Am. J. Cardiol. *59:* 72A–75A (1987).

Loutzenhiser, R.; Epstein, M.; Horton, C.; Sonke, P.: Reversal by the calcium antagonist nisoldipine of norepinephrine-induced reduction of GFR: evidence for preferential antagonism of preglomerular vasoconstriction. J. Pharmac. exp. Ther. *232:* 382–387 (1985).

Luft, F.C.; Aronoff, G.R.; Sloan, R.S.; Fineberg, N.S.; Weinberger, M.H.: Calcium channel blockade with nitrendipine. Hypertension *7:* 438–442 (1985).

Lundberg, J.M.; Hua, X.Y.; Franco-Cereceda, A.: Effects of NPY on mechanical activity and neurotransmission in the heart, vas deferens and urinary bladder of the guinea pig. Acta physiol. scand. *121:* 325–332 (1983).

Lundin, S.A.; Hallbäck-Nordlander, M.I.L.: Regression of structural cardiovascular changes by antihypertensive therapy in spontaneously hypertensive rats. J. Hypertens. *2:* 11–18 (1984).

Lund-Johansen, P.: Hemodynamic effects of verapamil in essential hypertension at rest and during exercise. Acta med. scand., suppl. 681, pp. 109–115 (1983a).

Lund-Johansen, P.: Hemodynamic response: decrease in cardiac output versus reduction in vascular resistance. Hypertension *5:* suppl. III, pp. 49–57 (1983b).

Mabe, Y.; Tatemoto, K.; Huidobro-Toto, J.P.: Neuropeptide Y-induced pressor responses: activation of a non-adrenergic mechanism, potentiation by reserpine and blockade by nifedipine. Eur. J. Pharmacol. *116:* 33–39 (1985).

Mace, P.J.; Stallard, T.J.; Littler, W.A.: The effect of felodipine on the forearm hemodynamics and the myogenic response of the forearm resistance vessels in normal man. Br. J. clin. Pharmacol. *20:* 383–386 (1985).

Magometschnigg, D.; Hortnagl, H.; Rameis, H.: Diltiazem and verapamil: functional antagonism of exogenous norepinephrine and angiotensin II in man. Eur. J. clin. Pharmacol. *26:* 303–307 (1984).

Majid, P. A.; De Jong, J.: Acute hemodynamic effects of nifedipine in patients with ischemic heart disease. Circulation *65:* 1114–1118 (1982).

Marre, M.; Misumi, J.; Raemsch, K.-D.; Corvol, P.; Menard, J.: Diuretic and natriuretic effects of nifedipine on isolated perfused rat kidneys. J. Pharmac. exp. Ther. *223:* 263–270 (1982).

Matsuzaki, M.; Gallagher, K. P.; Patritti, J.; Tajima, T.; Kemper, W. S.; White, F. C.; Ross, J. R.: Effects of a calcium entry blocker (diltiazem) on regional myocardial flow and function during exercise in conscious dogs. Circulation *69:* 801–814 (1984).

McDonagh, P. F.; Niven, A. T.; Roberts, D.: Direct visualization of the coronary microcirculation for pharmacologic and physiologic studies. Microvasc. Res. *28:* 180–196 (1984).

McGrath, J. C.; O'Brien, J. W.: Blockade by nifedipine of responses to intravenous bolus injection or infusion of alpha-1- and alpha-2-adrenoceptor agonists in the pithed rat. Br. J. Pharmacol. *91:* 355–365 (1987).

McLeay, R. A. B.; Stallard, T. J.; Watson, R. D. S.; Littler, W. A.: The effect of nifedipine on arterial pressure and reflex cardiac control. Circulation *67:* 1084–1090 (1983).

Morita, T.; Maniwa, T.; Satoh, K.; Taira, N.: Undifferentiated effects of calcium antagonists on pressor responses to selective alpha-1- and alpha-2-adrenoceptor agonists in anesthetized, spinal dogs. J. Pharmac. exp. Ther. *234:* 728–734 (1985).

Muiesan, G.; Agabiti-Rosei, E.; Castellano, M.; Alicandri, C. L.; Corea, L.; Fariello, R.; Meschi, M.; Romanelli, G.: Antihypertensive and humoral effects of verapamil and nifedipine in essential hypertension. J. cardiovasc. Pharmacol. *4:* suppl. 3, pp. 325–329 (1982).

Müller-Schweinitzer, E.: Tissue specific susceptibility of alpha-adrenoceptor mediated vasoconstriction to nifedipine. Arch. Pharmacol. *324:* 64–69 (1983).

Müller-Schweinitzer, E.; Neumann, E.: In vitro effects of calcium antagonists PN200-110, nifedipine, and nimodipine on human and canine cerebral arteries. J. cereb. Blood Flow Metab. *3:* 354–360 (1983).

Murphy, M. B.; Scriven, A. J.; Brown, M. J.; Causon, R.; Dollery, C. T.: The effects of nifedipine and hydralazine induced hypotension on sympathetic activity. Eur. J. clin. Pharmacol. *23:* 479–482 (1982).

Nagao, T.; Yamaguchi, I.; Narita, H.; Nakajima, H.: Calcium entry blockers: antihypertensive and natriuretic effects in experimental animals. Am. J. Cardiol. *56:* 56H–61H (1985).

Nakayama, H.; Schwartz, A.; Millard, R. W.: Reflex chronotropic and inotropic effects of calcium channel blocking agents in conscious dogs. Circulation Res. *52:* 302–311 (1983).

Nannan, M. E.; Melin, J. A.; Vanbutsele, R. J.; Lavenne, F.; Detry, J. M.: Acute and long-term effects of nitrendipine on resting and exercise hemodynamics in essential hypertension. J. cardiovasc. Pharmacol. *6:* S1043–S1048 (1984).

Narita, H.; Nagao, T.; Yabana, H.; Yamaguchi, I.: Hypotensive and diuretic actions of diltiazem in spontaneously hypertensive and Wistar-Kyoto rats. J. Pharmac. exp. Ther. *227:* 472–477 (1983).

Natsume, T.; Gallo, A.J.; Pegram, B.L.; Frohlich, E.D.: Hemodynamic effects of prolonged treatment with diltiazem in conscious normotensive and spontaneously hypertensive rats. Clin. exp. Hypertens. *A7:* 1471–1479 (1985).

Navar, L.G.; Champion, W.J.; Thomas, C.E.: Effects of calcium channel blockade on renal vascular resistance responses to changes in perfusion pressure and angiotensin-converting enzyme inhibition in dogs. Circulation Res. *58:* 874–881 (1986).

Nayler, W.G.; Dillon, J.S.: Calcium antagonists and their mode of action: an historical overview. Br. J. clin. Pharmacol. *21:* suppl., pp. 97–107 (1986).

Nayler, W.G.; Horowitz, J.D.: Calcium antagonists: a new class of drugs. Pharmacol. Ther. *20:* 203–262 (1983).

Nievelstein, H.M.N.W.; van Essen, H.; Tijssen, C.M.; Smits, J.F.M.; Struyker-Boudier, H.A.J.: Systemic and regional actions of calcium entry blockers in conscious spontaneously hypertensive rats. Eur. J. Pharmacol. *113:* 187–198 (1985).

Nilsson, J.; Sjölund, M.; Palmberg, L.; Van Euler, A.M.; Jonzon, B.; Thyberg, J.: The calcium antagonist nifedipine inhibits arterial smooth muscle cell proliferation. Atherosclerosis *58:* 109–122 (1985).

Nordlander, M.: Hemodynamic effects of short- and long-term administration of felodipine in spontaneously hypertensive rats. Drugs *29:* suppl. 2, pp. 90–101 (1985).

Nordlander, M.; Di Bona, G.F.; Ljung, B.; Yao, T.; Thorēn, P.: Renal and cardiovascular effects of acute and chronic administration of felodipine to SHR. Eur. J. Pharmacol. *113:* 25–36 (1985).

Nyborg, N.C.B.; Byg-Hansen, J.; Mulvany, M.J.: Effect of felodipine on resistance vessels from spontaneously hypertensive and normotensive rats. J. cardiovasc. Pharmacol. *7:* suppl. 6, pp. 43–46 (1985).

Nyborg, N.C.B.; Mulvany, M.J.: Effect of felodipine, a new dihydropyridine, on contractile responses to potassium, noradrenaline and calcium in mesenteric resistance vessels of the rat. J. cardiovasc. Pharmacol. *6:* 499–505 (1984).

Nyborg, N.C.B.; Mulvany, M.J.: Lack of effect of antihypertensive treatment with felodipine on cardiovascular structure of young spontaneously hypertensive rats. Cardiovasc. Res. *19:* 528–536 (1985).

Ogawa, N.; Ono, H.: Different effects of noradrenaline, angiotensin II and Bay K 8644 on the abolition of autoregulation of renal blood flow by verapamil. Arch. Pharmacol. *333:* 445–450 (1986).

Ogilvie, R.I.: Comparative effects of vasodilator drugs on flow distribution and venous return. Can. J. Physiol. Pharmacol. *63:* 1345–1355 (1985).

Olivari, M.T.; Bartorelli, C.; Polese, A.; Fiorentini, C.; Moruzzi, P.; Guazzi, M.D.: Treatment of hypertension with nifedipine, a calcium antagonistic agent. Circulation *59:* 1056–1062 (1979).

Pasanisi, F.; Elliott, H.L.; Meredith, P.A.; Sumner, D.J.; Reid, J.L.: Effect of calcium channel blockers on adrenergic and non-adrenergic vascular responses in man. J. cardiovasc. Pharmacol. *7:* 1166–1170 (1985).

Peach, M.J.; Singer, H.A.; Izzo, N.J.; Loeb, A.L.: The role of calcium in endothelium-dependent relaxation of arterial smooth muscle. Am. J. Cardiol. *59:* 35A–43A (1987).

Pedrinelli, R.; Fouad, F.M.; Tarazi, R.C.; Bravo, E.L.; Textor, S.C.: Nitrendipine, a calcium entry blocker. Archs. int. Med. *146:* 62–65 (1986).

Pedrinelli, R.; Tarazi, R.C.: Calcium entry blockade by nitrendipine and alpha-adrenergic responsiveness in vivo: comparison with noncalcium entry blocker vasodilators in absence and presence of phenoxybenzamine pretreatment. J. Pharmac. exp. Ther. *233:* 636–642 (1985a).

Pedrinelli, R.; Tarazi, R.C.: Calcium entry blockade by nitrendipine and alpha-adrenergic responsiveness in vivo: comparison of systemic vs. local effects. J. Pharmac. exp. Ther. *233:* 643–649 (1985b).

Pegram, B.L.; Kobrin, I.; Sesoko, S.; Frohlich, E.D.: Nitrendipine: hemodynamic effects in conscious normotensive and spontaneously hypertensive rats. J. cardiovasc. Pharmacol. *6:* suppl. 7, pp. 1016–1026 (1984).

Pelayo, J.C.: Modulation of renal adrenergic effector mechanisms by calcium entry blockers. Am. J. Physiol. *252:* F613–F620 (1987).

Pernow, J.; Lundberg, J.M.: Neuropeptide Y constricts human skeletal muscle arteries via a nifedipine-sensitive mechanism independent of extracellular calcium? Acta physiol. scand. *128:* 655–656 (1986).

Pernow, J.; Svenberg, T.; Lundberg, J.M.: Actions of calcium antagonists on pre- and postjunctional effects of neuropeptide Y on human peripheral blood vessels in vitro. Eur. J. Pharmacol. *136:* 207–218 (1987).

Rapoport, R.M.; Draznin, M.B.; Murad, F.: Mechanism of the adenosine triphosphate, thrombin, and trypsin-induced relaxation of the rat thoracic aorta. Circulation Res. *55:* 468–479 (1984).

Reed, B.V.; Tuma, R.F.: The effects of nifedipine on regional vascular resistance and blood flow in the spontaneously hypertensive rat. Clin. exp. Hypertens. *A8:* 963–979 (1986).

Reuter, H.: Calcium channel modulation by neurotransmitters, enzymes and drugs. Nature *301:* 569–574 (1983).

Rimele, T.J.; Rookw, T.W.; Aarhua, L.L.; Vanhoutte, P.M.: Alpha-1-adrenoceptors and calcium in isolated canine coronary arteries. J. Pharmac. exp. Ther. *226:* 668–672 (1983).

Robinson, B.F.; Collier, J.G.; Dobbs, R.J.: Comparative dilator effect of verapamil and sodium nitroprusside in forearm arterial bed and dorsal hand veins in man: functional differences between vascular smooth muscle in arterioles and veins. Cardiovasc. Res. *13:* 16–21 (1979).

Rousseau, M.F.; Etienne, J.; Van Mechelen, H.; Veriter, C.; Pouleur, H.: Hemodynamic and cardiac effects of nicardipine in patients with coronary artery disease. J. cardiovasc. Pharmacol. *6:* 833–839 (1984).

Rowland, E.; Evans, T.; Krikler, D.: Effect of nifedipine on atrioventricular conduction as compared with verapamil. Intracardiac electrophysiological study. Br. Heart J. *42:* 124–127 (1979).

Ruffolo, R.R.; Morgan, E.L.; Messick, K.: Possible relationship between receptor reserve and the differential antagonism of alpha-1- and alpha-2-adrenoceptor mediated pressor responses by calcium channel antagonists in the pithed rat. J. Pharmac. exp. Ther. *230:* 587–594 (1984).

Ruffolo, R.R.; Yaden, E.L.: Existence of spare alpha-1-adrenoceptors, but not alpha-2-adrenoceptors for respective vasopressor effects of cirrazoline and BHT-933 in the pithed rat. J. cardiovasc. Pharmacol. *6:* 1011–1019 (1984).

Ruskoaho, H.J.; Savolainen, E.R.: Effects of long-term verapamil treatment on blood pressure, cardiac hypertrophy and collagen metabolism in spontaneously hypertensive rats. Cardiovasc. Res. *19:* 355–362 (1985).

Rutsch, W.; Schmutzler, H.: Comparison of the acute hemodynamic effects of nifedipine with nitrendipine and a study of the electrophysiological effects of nitrendipine in man. J. cardiovasc. Pharmacol. *6:* suppl. 7, pp. 1011–1015 (1984).

Ryman, K.S.; Kubo, S.H.; Shaknovich, A.; Cody, R.J.: Influence of baseline hemodynamic status and sympathetic activity on the response to nicardipine, a new dihydropyridine, in patients with hypertension or chronic congestive heart failure. Clin. Pharmacol. Ther. *41:* 483–489 (1987).

Saeed, M.; Holtz, J.; Elsner, D.; Bassenge, E.: Attenuation of sympathetic vasoconstriction by nifedipine: the role of vascular alpha-2-adrenoceptors. Eur. J. Pharmacol. *94:* 149–154 (1983).

Safar, M.E.; Simon, A.Ch.; Levenson, J.A.; Cazor, J.L.: Hemodynamic effects of diltiazem in hypertension. Circulation Res. *52:* I169–I173 (1983).

Saida, K.; van Breemen, C.: Mechanism of Ca^{++} antagonist-induced vasodilation: intracellular actions. Circulation Res. *52:* 137–142 (1983).

Saito, I.; Takeshita, E.; Saruta, T.; Nagano, S.; Sekihara, T.: Effect of a calcium entry blocker on blood pressure, plasma renin activity, aldosterone and catecholamines in normotensive subjects. Clin. Endocrinol. *24:* 565–570 (1986).

Scarborough, N.L.; Carrier, G.O.: Nifedipine and alpha-adrenoceptors in rat aorta. The role of extracellular calcium in alpha-1- and alpha-2-adrenoceptor mediated contraction. J. Pharmac. exp. Ther. *231:* 597–602 (1984).

Schumann, H.J.; Gorlitz, B.D.; Wagner, J.: Influence of papaverine, D600, and nifedipine on the effects of noradrenaline and calcium on the isolated aorta and mesenteric artery of the rabbit. Arch. Pharmacol. *289:* 409–418 (1975).

Schwartz, J.S.; Bache, R.J.: Pharmacologic vasodilators in the coronary circulation. Circulation *75:* suppl. I, pp. 162–167 (1987).

Segstro, R.; Seaman, K.L.; Innes, I.R.; Greenway, C.V.: Effects of nifedipine on hepatic blood volume in cats: indirect venoconstriction and absence of inhibition of postsynaptic alpha-2-adrenoceptor responses. Can. J. Physiol. Pharmacol. *64:* 615–620 (1986).

Silke, B.; Frais, M.A.; Muller, P.; Verma, S.P.; Reynolds, G.; Taylor, S.H.: Hemodynamic dose-response effects of intravenous nisoldipine in coronary artery disease. Br. J. clin. Pharmacol. *20:* 675–680 (1985).

Silke, B.; Sharma, S.K.; Verma, S.P.; Midtbo, K.A.; Reynolds, G.; Taylor, S.H.: A hemodynamic and radionuclide assessment of diltiazem in coronary heart disease. Br. J. clin. Pharmacol. *23:* 165–172 (1987).

Silke, B.; Verma, S.P.; Nelson, G.I.; Hussain, M.; Taylor, S.H.: Hemodynamic dose-response effects of i.v. nicardipine in coronary artery disease. Br. J. clin. Pharmacol. *18:* 717–724 (1984).

Simon, G.; Snyder, D.K.: Altered pressor responses in long-term nitrendipine treatment. Clin. Pharmacol. Ther. *36:* 315–319 (1984).

Singer, H.A.; Peach, M.J.: Calcium and endothelial mediated vascular smooth muscle relaxation in rabbit aorta. Hypertension *4:* suppl. II, pp. II19–II25 (1982).

Singh, B.N.: The mechanism of action of calcium antagonists relative to their clinical applications. Br. J. clin. Pharmacol. *21:* suppl., pp. 109–121 (1986).

Sluiter, H.E.; Huysmans, F.T.; Thien, Th.A.; Van Lier, H.J.; Koene, R.A.: Hemodynamic, hormonal, and diuretic effects of felodipine in healthy normotensive volunteers. Drugs *29:* suppl. 2, pp. 26–35 (1985).

Smith, S.A.; Mac, P.J.; Littler, W.A.: Felodipine, blood pressure, and cardiovascular reflexes in hypertensive humans. Hypertension *8:* 1172–1178 (1986).

Smits, J.F.M.; Coleman, T.G.; Smith, T.L.; Kasbergen, C.M.; van Essen, H.; Struyker-Boudier, H.A.J.: Antihypertensive effect of propranolol in conscious spontaneously hypertensive rats: central hemodynamics, plasma volume and renal function during beta-blockade with propranolol. J. cardiovasc. Pharmacol. *4:* 903–914 (1982).

Sorkin, E.M.; Clissold, S.P.: Nicardipine: a review of its pharmacodynamic and pharmacokinetic properties, and therapeutic efficacy in the treatment of angina pectoris, hypertension and related cardiovascular disorders. Drugs *33:* 296–345 (1987).

Spies, H.F.; Koch, M.; Appel, E.; Palm, D.; Kaltenbach, M.M.: Einfluß oraler Langzeittherapie mit Verapamil und Metoprolol auf arteriellen Blutdruck, Herzfrequenz und Plasmakatecholamine bei Hypertonikern. Z. Kardiol. *67:* 196 (1978).

Stasch, J.-P.; Kazda, S.; Hirth, C.: Effect on hypertension, cardiac hypertrophy and atrial natriuretic peptides of treatment with nitrendipine in SHR. J. Hypertens. *4:* suppl. 6, pp. 160–162 (1986).

Steele, T.H.; Challoner-Hue, L.: Response of isolated Dahl rat kidney to calcium antagonists. Kidney int. *31:* 941–946 (1987).

Stevens, M.J.; Moulds, R.F.W.: Antagonism by nifedipine of alpha-1- and alpha-2-adrenoceptor-mediated responses of human digital arteries. J. Pharmac. exp. Ther. *236:* 764–768 (1986).

Stoepel, K.; Heise, A.; Kazda, S.: Pharmacological studies of the antihypertensive effect of nitrendipine. Arzneimittel-Forsch. *31:* 2056–2061 (1981).

Struyker-Boudier, H.A.J.: Dynamic systems analysis as a basis for drug design: application to antihypertensive drug therapy; in Ariens, Drug design, vol. X, pp.145–196 (Academic Press, New York, 1980).

Struyker-Boudier, H.A.J.: The interaction of antihypertensive drugs with mechanisms of blood pressure regulation; in Van Zwieten, Handbook of hypertension, vol. 3, pp. 46–65 (Elsevier, Amsterdam, 1984).

Struyker-Boudier, H.A.J.; Smeets, G.; Brouwer, G.M.; Van Rossum, J.M.: Central nervous system alpha-adrenergic mechanisms and cardiovascular regulation in rats. Archs. int. Pharmacodyn. Thér. *213:* 285–293 (1975).

Struyker-Boudier, H.A.J.; Smits, J.F.M.; Van Essen, H.: The role of the baroreceptor reflexion in the cardiovascular effects of propranolol in the conscious SHR. Clin. Sci. *56:* 163–167 (1985).

Sunderrajan, S.; Reams, G.; Bauer, J.H.: Renal effects of diltiazem in primary hypertension. Hypertension *8:* 238–242 (1986).

Sybertz, E.J.; Vander Vliet, G.; Baum, Th.: Analysis of the vasoconstrictor responses to potassium depolarization and norepinephrine and their antagonism by differing classes of vasodilators in the perfused rat hindquarters. J. Pharmac. exp. Ther. *227:* 621–626 (1983).

Takanaka, T.; Asano, M.; Shiono, J.K.; Shibasaki, M.; Inagaki, O.: Cardiovascular pharmacology of nicardipine in animals. Br. J. clin. Pharmacol. *20:* suppl., pp. 7–22 (1985).

Takata, Y.; Yamashita, Y.; Takishita, S.; Fujishima, M.: Influence of the sympathetic nervous system and vasopressin on the blood pressure lowering effect of nifedipine in DOCA-salt hypertensive rats. Clin. Sci. *73:* 253–258 (1987).

Takeshita, A.; Mark, A. L.: Decreased vasodilator capacity at forearm vasculare resistance vessels in borderline hypertension. Hypertension *2:* 610–616 (1986).

Taylor, D. G.; Kowalski, T. E.: Comparison of calcium channel inhibitors on vagal heart rate responses elicited by arterial baroreceptor reflexes in anesthetized dogs. J. Pharmac. exp. Ther. *228:* 491–499 (1984).

Thuillez, C.; Gueret, M.; Duhaze, P.; Lhoste, F.; Kiechel, J. R.; Giudicelli, J. F.: Nicardipine: pharmacokinetics and effects on carotid and brachial blood flows in normal volunteers. Br. J. clin. Pharmacol. *18:* 837–847 (1984).

Timmermans, P. B. M. W. M.; Mathy, M. J.; Wilffert, B.; Kalkman, H. O.; Thoolen, M. J. M. C.; De Jonge, A.; Van Meel, J. C. A.; Van Zwieten, P. A.: Differential effect of calcium entry blockers on alpha-1-adrenoceptor-mediated vasoconstriction in vivo. Arch. Pharmacol. *324:* 239–245 (1983a).

Timmermans, P. B. M. W. M.; De Jonge, A.; Van Meel, J. C. A.; Mathy, M. J.; Van Zwieten, P. A.: Influence of nifedipine on functional responses in vivo initated at alpha-2-adrenoceptors. J. cardiovasc. Pharmacol. *5:* 1–11 (1983b).

Timmermans, P. B. M. W. M.; Mathy, M. J.; Thoolen, M. J. M. C.; De Jonge, A.; Wilffert, B.; Van Zwieten, P. A.: Invariable susceptibility to blockade by nifedipine of vasoconstriction to various alpha-2-adrenoceptor agonists in pithed rats. J. Pharmac. Pharmacol. *36:* 772–775 (1984).

Timmermans, P. B. M. W. M.; Thoolen, M. J. M. C.; Mathy, M. J.; Wilffert, B.; De Jonge, A.; Van Zwieten, P. A.: Effect of the irreversible alpha-adrenoceptor antagonists prenoxybenzamine and benextramine on the effectiveness of nifedipine in inhibiting alpha-1- and alpha-2-adrenoceptor-mediated vasoconstriction in pithed rats. Arch. Pharmacol. *329:* 404–413 (1985).

Towart, R.; Kazda, S.: The cellular mechanism of action of nimodipine (BAY e 9736), a new calcium antagonist. Br. J. Pharmacol. *67:* 409P–410P (1979).

Van Baak, M. A.; Struyker-Boudier, H. A. J.; Smits, J. F. M.: Antihypertensive mechanisms of beta-adrenoceptor blockade: a review. Clin. exp. Hypertens. *A7:* 1–72 (1985).

Van Breemen, C.; Farinas, B. R.; Gerba, P.; McNaughton, E. D.: Excitation contraction coupling in rabbit aorta studied by the lanthanum method for measuring cellular calcium influx. Circulation Res. *30:* 44–54 (1972).

Van Breemen, C.; Aaronson, P.; Loutzenhiser, P.: Na-Ca interactions in mammalian smooth muscle. Pharmac. Rev. *30:* 167–208 (1979).

Van Breemen, C.; Hwang, O.; Meisheri, R. D.: The mechanism of inhibitory action of diltiazem on vascular smooth muscle contractility. J. Pharmac. exp. Ther. *218:* 459–463 (1981).

Van Meel, J. C. A.; De Jonge, A.; Kalkman, H. O.; Wilffert, B.; Timmermans, P. B. M. W. M.; Van Zwieten, P. A.: Vascular smooth muscle contraction initiated by postsynaptic alpha-2-adrenoceptor activation is induced by an influx of extracellular calcium. Eur. J. Pharmacol. *69:* 205–208 (1981).

Van Meel, J. C. A.; De Zoeten, K.; Timmermans, P. B. M. W. M.; Van Zwieten, P. A.: Impairment by nifedipine of vasopressor responses to stimulation of postsynaptic alpha-2-adrenoreceptors in ganglion-blocked rabbits. Further evidence for the selective inhibition of postsynaptic alpha-2-adrenoceptor-induced pressor responses by calcium antagonists. J. autonom. Pharmacol. *2:* 13–20 (1982).

Van Meel, J. C. A.; Timmermans, P. B. M. W. M.; Van Zwieten, P. A.: Alpha-1- and alpha-2-

adrenoceptor stimulation in the isolated perfused hindquarters of the rat: an in vitro model. J. cardiovasc. Pharmacol. *5:* 580–585 (1983).

Van Schayk, B.A.M.; Van Nistelrooy, A.E.J.; Geijskes, G.G.: Antihypertensive and renal effects of nicardipine. Br. J. clin. Pharmacol. *18:* 57–63 (1984).

Vemulapalli, S.; Chiu, P.J.S.; Sybertz, E.J.: Comparative renal effects of calcium channel blockers in conscious spontaneously hypertensive rats. Archs int. Pharmacodyn. Thér. *287:* 309–322 (1987).

Ventura, H.O.; Messerli, F.H.; Oigman, W.; Dunn, F.G.; Reisin, E.; Frohlich, E.D.: Immediate hemodynamic effects of a new calcium channel blocking agent (nitrendipine) in essential hypertension. Am. J. Cardiol. *51:* 783–786 (1983).

Vogt, A.; Neuhaus, K.-L.; Kreuzer, H.: Hemodynamic effects of the new vasodilator drug Bay k 5552 in man. Arzneimittel-Forsch. *30:* 2162–2164 (1980).

Vollmer, R.R.; Fluharty, S.J.: Effects of 6-hydroxydopamine on hemodynamic function in the pithed rat. J. Pharmac. exp. Ther. *240:* 421–427 (1987).

Wallia, R.; Greenberg, A.; Puschett, J.B.: Renal hemodynamic and tubular transport effects of nitrendipine. J. Lab. clin. Med. *105:* 498–503 (1985).

Warltier, D.C.; Zyvoloski, M.G.; Gross, G.J.; Brooks, H.L.: Comparative actions of dihydropyridine slow channel calcium blocking agents in conscious dogs: alterations in baroreflex sensitivity. J. Pharmac. exp. Ther. *230:* 376–382 (1984).

Weinstein, D.B.; Heider, J.G.: Antiatherogenic properties of calcium antagonists. Am. J. Cardiol. *59:* 163B–172B (1987).

Wigley, F.M.; Wise, R.A.; Malamet, R.; Scott, T.E.: Nicardipine in the treatment of Raynaud's phenomenon. Arthritis Rheum. *30:* 281–287 (1987).

Williams, J.S.; Baik, Y.H.; Koch, W.J.; Schwartz, A.: A possible role for the endothelium in porcine coronary smooth muscle responses to dihydropyridine calcium channel modulators. J. Pharmac. exp. Ther. *241:* 379–386 (1987).

Winquist, R.J.; Bunting, P.B.; Schofield, T.L.: Blockade of endothelium-dependent relaxation by the amiloride analog dichlorobenzamil: possible role of Na^+/Ca^{2+} exchange in the release of the endothelium-derived relaxant factor. J. Pharmac. exp. Ther. *235:* 644–650 (1985).

Woodman, O.L.; Constantine, J.W.; Vatner, S.F.: Nifedipine attenuates both alpha-1- and alpha-2-adrenoceptor-mediated pressor and vasoconstrictor responses in conscious dogs and primates. J. Pharmac. exp. Ther. *239:* 648–653 (1986).

Yokoyama, S.; Kaburagi, T.: Clinical effects of intravenous nifedipine on renal function. J. cardiovasc. Pharmacol. *5:* 67–71 (1983).

Young, M.A.; Watson, R.D.S.; Stallard, T.J.; Littler, W.A.: Calcium channel blockers – are they diuretics? Br. J. clin. Pharmacol. *20:* suppl., pp. 95–98 (1985).

H.A.J. Struyker-Boudier, PhD, Department of Pharmacology,
University of Limburg, PO Box 616, NL–6200 MD Maastricht (The Netherlands)

Van Zwieten PA (ed): Clinical Aspects of Calcium Entry Blockers.
Prog Basic Clin Pharmacol. Basel, Karger, 1989, vol 2, pp 67–87

Calcium Antagonism in the Control of Cardiac Arrhythmias[1]

Bramah N. Singh

Department of Cardiology, Wadsworth Veterans Administration Medical Center, and Department of Medicine, UCLA School of Medicine, Los Angeles, Calif., USA

Since the description of the initial electrophysiologic actions of verapamil [1], the prototype of calcium antagonists (CAs), much evidence has accumulated for the role of CAs as potent agents for the control of certain supraventricular and ventricular arrhythmias [2]. Singh and Vaughan Williams [1] found that verapamil differed in its electrophysiologic properties from those of β-blocking drugs and of local-anesthetic type of antiarrhythmic compounds such as quinidine. The drug also differed from those agents (e.g., sotalol and amiodarone) which exerted their salutary effects by homogeneously prolonging the cardiac action potential and refractoriness without much effect on conduction.

The antiarrhythmic action of verapamil was therefore thought to constitute a discrete mechanism for the control of cardiac arrhythmias [1, 3]. Recent experimental and clinical studies have supported these earlier observations. The role of CAs as antiarrhythmic agents is discussed in this paper relative to the electrophysiologic properties of these compounds.

Electropharmacologic Considerations

Most arrhythmias arise from perturbations in impulse generation (ectopic or automatic tachyarrhythmias) and in impulse conduction (reentrant arrhythmias) or from a combination of both processes [4].

[1] Supported in part by the Medical Research Service of the Veterans Administration and the American Heart Association, the Greater Los Angeles, Affiliate, California.

Recently, "triggered" automaticity [5, 6] also has been shown to be a significant mechanism but the precise role of the abnormalities of the slow channel in the genesis of such arrhythmias remains essentially inferential.

In interpreting the actions of a calcium blockers as antiarrhythmic agents, it should be appreciated that the normal atrioventricular (A-V) node is slow channel dependent, a potential site of deranged impulse formation and impulse conduction. Calcium channel blockers block the slow-calcium channel in the A-V node in a concentration- or dose-dependent fashion, whereas in fast-channel fibers such as those in atria, ventricle or His-Purkinje system, they have no effect on refractory periods or on conduction velocity. In isolated preparations, CAs slowphase 4 depolarization in the sinoatrial and A-V nodes with an associated depression of conduction. Because of their effects on the slow channel, calcium channel blockers also accelerate repolarization at the plateau phase of the action potential. However, the magnitude of this effect is such that there is little or no change in the voltage-dependent refractoriness. The major effect on the refractory period is in the A-V node where both the effective and the functional refractory periods are significantly lengthened in the anterograde as well as in the retrograde directions. Some of the newer agents exert additional effects over and above their propensity to block the slow myocardial channel. For example, bepridil [7] blocks the fast-sodium channel and it prolongs the action potential duration but in concentrations higher than those depressing the slow channel.

Clinical Electrophysiologic Effects of Calcium Antagonists

The in vivo and clinical effects of CAs represent a balance of their in vitro actions and those that result from reflex activation of the sympathetic reflexes activated by their often potent vasodilator actions. In the case of nifedipine and other dihydropyridine derivatives, such reflex actions nullify or even reverse the intrinsic actions on the sinus and A-V nodes. Thus, these agents do not exert measurable effects on the electrophysiologic parameters in man and appear to be devoid of antiarrhythmic actions. In the case of diltiazem, verapamil and bepridil, the net effects are also modified by the noncompetitive sympatholytic effects of these compounds.

The clinical electrophysiologic actions relevant to the antiarrhythmic effects of various CAs are summarized in table I. The most significant effect is on the A-V node, an example of a "direct" antiarrhythmic action as exemplified in the termination of the acute episodes and for the prevention of recurrences of paroxysmal supraventricular tachycardia (PSVT). Most

Table I. Clinical electrophysiologic effects of CAs

Effects	Verapamil[1]	Nifedipine	Diltiazem	Bepridil
R-R interval				
QRS	0	0	0	↑
Q-Tc	0	0	0	↑
PR	↑↑	0	↑	↑
A-H	↑↑↑	0	↑↑	↑
H-V	0	0	0	↑
Atrial ERP	0	0	0	↑
A-V node ERP	↑↑↑	±	↑↑	↑↑
A-V node FRP	↑↑↑	±	↑↑	↑↑
Ventricular FRP	0	0	0	↑↑
His-Purkinje ERP	0	0	0	↑
Bypass tact ERP	±	0	±	↑
Sinus node recovery time	0[2]	0	0[2]	0
Ventricular automaticity	0	0	0	↓

↓ = Decrease; ↑ = increase; ± = variable effect; ERP = effective refractory period; FRP = functional refractory period.
[1]effects of gallopamil and tiapamil are similar to those of verapamil.
[2]prolonged in sick sinus syndrome.

CAs have little or no electrophysiologic effects on ventricular muscle. For this reason they are unlikely to be potent antiarrhythmic agents in most types of ventricular tachyarrhythmias, although recent experimental and clinical evidence has suggested a role for these drugs in the control of arrhythmias due to triggered automaticity. There is also evidence that CAs may prevent the development of such dysrhythmias by primarily influencing the course of myocardial ischemia either by altering demand or supply or both. The clearest example is the reversal of coronary vasospasm and control of vetricular tachyarrhythmias, an example of an "indirect" antiarrhythmic effect.

Effects of Calcium Antagonists in Supraventricular Arrhythmias

A simplified classification of the presumed electrophysiologic mechanism for the clinically occurring PSVTs is presented in table II. The automatic form is uncommon, especially in the adult; the commonest form

Table II. Mechanisms of PSVTs

Enhanced automaticity
Paroxysmal and acute
Chronic
Reentry without bypass tracts
A-V nodal reentry
Slow-fast
Fast-slow
Sinoatrial nodal reentry
Intra-atrial reentry
Reentry in association with bypass tracts
Reentry with anterograde A-V conduction (orthodromic)
With evidence of preexcitation of 12-lead ECG
Concealed Wolff-Parkinson-White syndrome (bypass tract conducting only retrogradely
Reentry with anterograde conduction over bypass tract (antidromic during tachycardia

is the reentrant PSVT [8]. PSVTs due to sinus nodal or intra-atrial reentry are uncommon; they are often difficult to distinguish from other types of reentrant PSVTs.

CAs have a role in the acute termination of PSVTs when given intravenously as well as in preventing recurrences when they are administered orally. Only those CAs which depress intranodal conduction and prolong the effective refractory period of the A-V node (verapamil, diltiazem, bepridil, tiapamil and gallopamil) are effective in this regard. The commonly used dosage regimens for the various agents are indicated in table III.

Acute Termination of Paroxysmal Supraventricular Tachycardia

The largest experience has been with intravenous verapamil but tiapamil, gallopamil, diltiazem and bepridil are also effective in terminating reentrant PSVTs: the locus of action is the anterograde limb of the tachycardia circuit in the A-V node, the "weakest link" of the reentrant loop. This accounts for the rapid and predictable reversion of 80–100% [9–13] of cases of PSVT given 3–5 mg (in children) to 10–15 mg (in adults) of verapamil.

Verapamil continues to be the drug of choice for terminating PSVTs with narrow QRS complexes due to reentry involving antegrade conduction over the A-V node. Diltiazem also has been shown recently to be effective;

Table III. Dosage regimens of CAs employed in the control of cardiac arrhythmias

Agent	Intravenous dose	Oral dose	Comments
Verapamil	0.145 mg/kg (bolus)	240–480 mg/day	regimen oral administered in 3 or 4 divided doses
	0.005 mg/kg/min (infusion)	–	
Diltiazem	0.15–0.25 mg/kg (bolus)	90–360 mg/day	oral regimen administered in 3 or 4 divided doses
Triapamil	1–2 mg/kg (bolus)	not established	–
Gallopamil (D_{600})	0.03–0.05 mg/kg (bolus)	not established	–
Bepridil	2–4 mg/kg	300–400 mg/day	oral regimen administered in one single daily dose

The dosages cited are well established for verapamil (the 'reference' CA). For the other compounds they are tentative and given to serve as an approximate indication.

intravenously (0.25 mg/kg) or orally it terminates narrow-QRS-complex PSVT [14–17], with a potency probably comparable to that of verapamil although a direct controlled comparison has not been made. The success rate of conversion of PSVTs to sinus rhythm by verapamil may be improved to nearly 100% by carotid sinus massage or the addition of 5–10 mg of edrophonium (Tensilon) given in rapid succession after verapamil. The drug is equally effective in terminating slow-fast as in fast-slow PSVTs. Verapamil is effective in terminating A-V nodal reentrant PSVTs at all ages. The major drawback of CA's is in patients with severe hypotension, heart failure or in antidromic PSVTs.

Modes of Conversion of Paroxysmal Supraventricular Tachycardia with Calcium Antagonists

The commonest is an abrupt termination of the tachyarrhythmia as might occur with carotid sinus massage. The cycle length of the tachycardia may prolong somewhat due to the delayed A-V conduction (as evidenced by lengthening of the A-H interval), followed by a short pause before conversion. The termination is usually by anterograde conduction block, rarely retrograde [12]. In some cases of PSVT given intravenous verapamil there is frank A-V dissociation followed by a junctional escape rhythm before sinus mechanism is restored. In some patients, restoration to sinus rhythm is preceded by transient atrial fibrillation. The underlying mechanism for this has not been defined. Another type of response includes the occurrence of premature ventricular contractions (PVCs; rarely short runs of ventricular tachycardia) preceding conversion to sinus rhythm. This is also seen during PSVT conversion by the use of β-adrenoceptor-blocking drugs or carotid sinus massage. Neither the mechanism nor the precise clinical significance of such PVCs is known. However, the occurrence of sustainded ventricular tachycardia complicating conversion of PSVT by verapamil is unusual.

The final type of response during PSVT termination is alternating cycle length preceding conversion [10]. Intracardiac recordings [13] have shown that cycle alternation is due to alterations in the A-H or H-A intervals but the underlying mechanism for this phenomenon is not understood: however, it is not a unique feature of the action of CAs [see ref. 10].

Acute Conversion of Paroxysmal Supraventricular Tachycardia: Calcium Antagonists versus Other Antiarrhythmic Agents

Intravenous β-blocking drugs produce a conversion rate of 40–60% of all cases of reentrant PSVT with narrow QRS complexes. There are few systematic data on the overall efficacy of vagal maneuvers, intravenous edrophonium, α-agonists or digoxin given either alone or in combination. However, verapamil is now preferred where urgent termination of PSVT is desirable since the effects of digoxin are often not apparent for 15–60 min or longer. In the rare event of failure of conversion with verapamil, DC conversion may be undertaken reasonably promptly because of the short elimination half-life of intravenous verapamil (30 min). In addition to the higher conversion rate, verapamil also offers other advantages over β-blockers. For example, in patients whose tachyarrhythmias are associated with chronic obstructive airways disease, diabetes mellitus or peripheral vascular disease, it is better tolerated than β-blockers.

It is also of interest that Class I agents such as quinidine [18], disopyramide [19] and ajmaline [20] may also convert PSVT acutely when given either orally or intravenously. All these three compounds, which also slow conduction and prolong refractoriness over the bypass tracts in the Wolff-Parkinson-White syndrome, prevent A-V nodal reentry by the selective block of the retrograde fast intranodal pathway. Recent experience has indicated that Class Ic agents such as flecainide, encainide and propafenone may also be effective in this regard. The overall merit of Class I agents versus that of verapamil and diltiazem is poorly defined.

Calcium Antagonists versus Adenosine and Adenosine Triphosphate in the Acute Conversion of Paroxysmal Supraventricular Tachycardia

Recent reports [21] have suggested that intravenously administered adenosine or adenosine triphosphate (ATP) may not only be as effective as verapamil but may also offer other important advantages [21, 22] at least in certain subsets of patients. However, at present this approach is not an approved indication in the USA and many other countries. Numerous studies with intravenous ATP (3–15 mg in pediatric patients and 10–20 mg in adults) and adenosine (mean dose of 180 μ/kg) given as bolus injections have nevertheless indicated that between 90 and 95% of PSVTs are terminated within 40 s). The conversion may be attended by transient (lasting few seconds) sinus arrest, A-V block, bradycardia, facial flushing, bronchospasm and even convulsions in the case of ATP. The side effects with adenosine appear to be minimal [21] but those of ATP are greater than those of intravenous verapamil [22]. Further studies are needed, especially those using blinded protocols comparing the efficacy of intravenous verapamil and adenosine.

Response in Other Forms of Reentrant of Ectopic Supraventricular Tachycardia

Verapamil and similar CAs may be of value in other forms of reentrant and automatic supraventricular tachycardia [10]. However, it appears that intravenous verapamil is less effective in converting ectopic atrial tachycardia [13] although paroxysmal atrial tachycardia with block due to digoxin toxicity, as well as to other causes, may convert to sinus rhythm following the oral doses of verapamil. The mechanism mediating such a beneficial response is not known; it may be due to the drug's effect on digitalis-induced triggered automaticity [23].

Calcium Antagonist and Multifocal Atrial Tachycardia

The role of CAs in multifocal atrial tachycardia used for acute tretment or prophylactically, is still uncertain. In some studies, beneficial effects have been reported; in others less encouraging effects have been described. For example, Aronow et al. [24] compared responses of verapamil in reentrant PSVT to those to multifocal atrial tachycardia; of the 5 patients with reentrant PSVT, all converted to sinus rhythm after intravenous verapamil, whereas, only 1 of 5 with multifocal atrial tachycardia responded. In contrast, a recent study by Levine et al. [25] indicated a generally favorable response. They found that intravenous verapamil consistently reduced the atrial and ventricular rates in a group of patients with multifocal atrial tachycardia and in 1 patient chronically administered drug prevented the recurrences of the tachycardia. Further data are needed to deny or confirm such a beneficial response.

Chronic Prophylaxis of Paroxysmal Supraventricular Tachycardia with Calcium Antagonists

Again, the bulk of experience has been with verapamil but even in the case of verapamil, available data [13, 26–29] suggest that, in contrast to the extreme potency of the intravenous drug, the oral formulation (given 80–120 mg tid or qid) may not have comparable efficacy in preventing recurrences of PSVT during chronic administration. However, Mauritson et al. [26] found that during long-term therapy of PSVT, oral verapamil was superior to placebo. Rinkenberger et al. [13] found that when the drug was given orally (180–480 mg/day) to patients who had responded to intravenous verapamil during acute episodes of PSVT, 10 of 19 patients discontinued therapy within a month, either because of side effects or lack of response.

Because of the extremely variable pattern of recurrences of PSVT in different patients and in the same patient at different times, the precise efficacy of a therapeutic agent in the chronic prophylaxis of PSVT is not easily established. Nevertheless, in the case of verapamil, there are reports which indicate that this compound may be of prophylactic value in PSVT [29].

An approach that appears to be effective in predicting the efficacy of oral CAs in the prophylaxis of PSVT is the use of electrophysiologic studies by programmed electrical stimulation [30, 31]. Tonkin et al. [30] made observations before and after intravenous verapamil (0.15 mg/kg), in 13 patients, 12 previously resistant to other antiarrhythmic drugs. The drug increased the A-V nodal conduction time as well as the functional and the effective refractory periods. After the drug, reentrant PSVT could not be initiated in 5 of the 13 patients and was nonsustained in a further 3 patients. The tachycardia rate was decreased in those in whom PSVT was reinducible after intravenous verapamil. Over a mean follow-up period of 16 months on oral verapamil, 11 of the 13 patients had definite symptomatic improvement, with a decrease in the frequency, duration and/or associated symptoms of their tachyarrhythmia. In only 1 patient were side effects severe enough to necessitate withdrawal of the drug. Very similar results have been reported recently in 8 patients by Klein et al. [31]. Their data suggested that the response to intravenous verapamil could be extrapolated to predict the electrophysiologic results and the outcome of long-term therapy with the oral drug; a similar predictive outcome between the acute intravenous effect and the chronic clinical response in 18 patients given diltiazem has been reported by Hung et al. [16]. However, more extensive controlled experiences will be needed to define clearly the role of oral verapamil, diltiazem, and other agents in the prophylaxis of PSVT.

There is a rational basis for modulating the electrophysiologic properties of the A-V node by combination therapy with β-adrenoceptor-blocking drugs or digoxin; such an approach may enhance the clinical utility of verapamil and other rate-lowering CAs (diltiazem, bepridil) in the oral prophylaxis of PSVT.

Verapamil and Paroxysmal Atrial Tachycardia with Block

Paroxysmal atrial tachycardia with block is a relatively uncommon arrhythmia, believed, in most cases to be due to digitalis intoxication. However, in the experience of Storstein and Landmark [32], digitalis could not be incriminated as a cause in many of their cases. They found that verapamil given orally in a dose of 40–80 mg 3-hourly to 14 patients with paroxysmal atrial tachycardia with block led to reversion to sinus rhythm in 10 initially, with subsequent relapse occuring in 4. The best effect was found in patients with a short duration of the arrhythmia. The mechanism of action of verapamil in this setting is uncertain but the data suggest that the drug has a depressant effect which may reflect actions on triggered automaticity induced by cardiac glycosides.

Calcium Antagonists in Preexcitation Syndrome

The major action of CAs in preexcitation syndromes is on the A-V node [10]. There is a minimal effect on the bypass tracts. Thus, compounds such as verapamil, tiapamil, gallopamil, diltiazem and bepridil are likely to be variably effective in terminating the acute paroxysms of the narrow-QRS orthodromic PSVT. As in the case of A-V nodal reentrant PSVT, the role of CAs in the oral prophylaxis of orthodromic PSVT remains to be defined but, since they prolong A-V nodal refractoriness, they are likely to be of value at least in some patients. In contrast, since they have no significant effect on the refractoriness of the bypass tract, they are unlikely to terminate antidromic PSVT; nevertheless, the theoretical possibility exists that they may, in large doses, terminate the arrhythmias by inhibiting retrograde A-V nodal conduction [10].

In patients with atrial flutter and fibrillation complicating preexcitation, agents which shorten the effective refractory period of the bypass tract (e.g., digitalis) and those which lengthen it over the A-V node (digitalis and β-adrenoceptor-blocking drugs) will augment the ventricular response and may possibly precipitate ventricular fibrillation. CAs which increase A-V nodal refractoriness (e.g., verapamil, diltiazem, tiapamil, gallopamil and bepridil) fall into the latter category. Published data, however, pertain largely to verapamil [see ref. 10]. They indicate that verapamil and similar CAs should not be given to patients in whom atrial fibrillation or flutter complicates the Wolff-Parkinson-White syndrome with anterograde conduction over the bypass tracts. These conditions are best treated acutely by cardioversion or class I drugs (procainamide, disopyramide, encainide or flecainide) and prophylactically by amiodarone or quinidine or possibly surgery. It should be emphasized that when doubt exists regarding the nature of wide-QRS tachycardia (irregular or regular), it is prudent not to use CAs for acute conversion.

Treatment of Other Atrial Tachyarrhythmias by Calcium Antagonists

Atrial Fibrillation

In atrila fibrillation, three types of responses to intravenous verapamil have been reported. The commonest is the inhibition of A-V conduction with subsequent slowing of the ventricular response [10]. The effect is short lived; after 30 min the ventricular response gradually accelerates, unless a continuous infusion is instituted. The second type of response is what has been termed "regularization" of the ventricular response. This occurred in 25% in the series of Heng et al. [9] but in a higher proportion of those reported by Schamroth et al. [11]. The precise mechanism of the phenomenon is unknown. The final type of response of atrial fibrillation to intravenous verapamil is reversion to sinus rhythm. This is, however, uncommon. It usually occurs in patients with arrhythmia of recent onset associated with normal, or only mild to moderate, left atrial enlargement. This observation is in line with the known findings that verapamil and similar drugs have no significant effect on atrial refractoriness [9].

The clinical value of orally administered verapamil in controlling ventricular response in atrial fibrillation has recently been studied by a number of investigators [10]. In one study, the effects of verapamil

(160–240 mg daily) on resting heart rate and that during mild exercise were studied in 23 digitalized patients with chronic atrial fibrillation of various causes. Verapamil significantly reduced the heart rate response during exercise; in a smaller number of patients another study [33] also confirmed the efficacy of verapamil in attenuating the ventricular response in atrial fibrillation. However, in this study the higher doses were associated with significant side effects. It should be noted that verapamil tends to reduce both the resting as well as the exercise-induced increases in heart rate in the setting of atrial fibrillation, whereas the major effect of digoxin is on the resting rate. Two recent reports are of interest [34, 35]. The first deals with the effects of varying doses of verapamil in the control of the ventricular response in atrial flutter and fibrillation. Klein and Kaplinsky [34] have shown clearly that not only does the CA effectively control the ventricular response as well as digitalis, but during exercise the attenuation of the rate increase is superior. They have raised thc issue that verapamil might supplant cardiac glycosides in this setting. Roth et al. [35] present similar data in the case of diltiazem. It is likely that the effects of the two drugs are due not solely to calcium antagonism but may in part arise from their associated noncompetitive sympatholytic effects. It should be emphasized that the suggestion of Klein and Kaplinsky [34] is essentially valid for most patients, especially those with preserved ventricular function. In patients with impaired ventricular and atrial function the effects of verapamil and diltiazem have not been evaluated critically and, at least for the present, cardiac glycosides may be preferred.

From the pharmacologic standpoint, it is clear that there is likely to be a subset of patients in whom digoxin and verapamil (or diltiazem) might be combined in the control of the ventricular response in atrial flutter and fibrillation. However, it should be emphasized that when digoxin and verapamil are used together their depressant effect on the A-V node may summate; since a significant interaction occurs between the two compounds the net effect may be unexpectedly exaggerated, necessitating careful monitoring of serum drug concentrations. Although an interaction between diltiazem and digoxin is either minimal or absent, the electrophysiologic interaction between the two drugs on the A-V node may produce a marked depressant effect in patients with unsuspected conduction system disease.

It is clear that since verapamil and similar compounds do not exacerbate bronchospastic or vasospastic phenomena, if effective in atrial tachyarrhythmias in patients with obstructive airways or peripheral vascular disease, these drugs will have distinct advantages over β-antagonists for the purposes of chronic prophylaxis in the above subsets of patients.

Maintenance of Sinus Rhythm

Basu et al. [36] reported that patients with atrial fibrillation treated with verapamil before and after DC cardioversion had a low rate of conversion (1 out of 24) *before* electroconversion was attempted. In contrast, on verapamil low energy levels were needed for conversion to sinus rhythm in the remainder, 40% of whom remained in sinus rhythm on the drug at the end of 1 year. The reported success rate with quinidine is 18–29%. In contrast, a recent controlled clinical trial [37] comparing verapamil (80 mg t.i.d.) with quinidine (0.4 mg b.i.d.) demonstrated that quinidine was significantly more effective in producing conversion to sinus rhythm at 3 months after conversion.

Atrial Flutter

The bulk of the clinical experience has been with verapamil [10]. The immediate effect of intravenous verapamil is reasonably consistent; the increased A-V block slows the ventricular response, with an occasional patient converting to sinus rhythm. In some, atrial flutter alters to atrial fibrillation before sinus rhythm is restored but the rate of conversion to sinus rhythm is low, which is consistent with the known electrophysiologic effects of the drug.

A single intravenous dose of verapamil has been found to be of diagnostic value in differentiating atrial flutter with 2-1 A-V conduction from PSVT when these two arrhythmias are not readily distinguished electrocardiographically. If the rhythm is atrial flutter, the A-V block increases immediately after intravenous verapamil, thus revealing the true nature of the arrhythmias [10]. Other CAs which inhibit A-V conduction may also be used in this way.

Ventricular Arrhythmias

Unlike the situation with supraventricular tachyarrhythmias as outlined above, the role of CA in ventricular dysrhythmias is poorly defined. In general, these agents are of limited value in most patients with life-threatening ventricular arrhythmias. The subject recently has been placed in perspective by Belhassen and Horowitz [38]. There are theoretical considerations which suggest that CAs might be of value in torsades de pointes (see below). It must be indicated that there is increasing evidence to implicate the potential role of calcium ion in early ischemic and reperfusion arrhythmias [23]. The data are consistent with the early observations of Kaumann and Aramendia [41] that pretreatment with verapamil exerted a potent antifibrillatory effect following coronary artery ligation in dogs. At present, it is unclear how these observations in experimental animals may be correlated with the clinical experience the bulk of which indicates a limited role of CAs in clinically occurring ventricular arrhythmias, except in well-defined small subsets of patients. The effects of these drugs in different forms of ventricular arrhythmias are discussed below.

Premature Ventricular Contractions

There are no controlled data to support the belief that CAs are potent agents for the suppression of PVCs in the setting of chronic ischemic heart disease. Thus, if they were to prevent the incidence of sudden death in such patients, the mechanism of the salutary effect is unlikely to be mediated through the suppression of PVCs. Similarly, while reasonably sound theoretical bases have been suggested for the role of the slow channel in the genesis of ventricular arrhythmias in obstructive cardiomyopathies [39] and for the mitral valve syndrome [2,5], clinical experience with slow-channel inhibitors in these conditions has not confirmed their efficacy, especially in obstructive cardiomyopathies [40]. Moreover, there are few systematic data to indicate that they may be effective in reducing the frequency of symptomatic PVCs in patients without structural heart disease.

Chronic Recurrent Ventricular Tachycardia

Verapamil and other CAs have little or no effect on chronically occurring ventricular tachyarrhythmias [38, 42, 43]. This is consistent with the fact that the effective refractory period of the ventricle or His Purkinje system is not lengthened by slow-channel blockade [41]. Wellens et al. [42] studied the effects of intravenous verapamil on ventricular tachycardia (VT) in patients in whom VT could be induced and terminated by programmed electrical stimulation of the heart. Intravenous verapamil (10 mg) given during the tachycardia had no effect in 7 of 8 patients, and the arrhythmia could be induced readily before as well as after verapamil administration. In only 1 patient could the induced VT be terminated by verapamil administration.

Sung et al. [44], who have reported similar findings, gave intravenous verapamil to 10 patients with inducible VT (8 with programmed stimulation and in 2 with isoproterenol). The drug had no effect on VT induction in any of the 10 patients. Cardioversion was required in 2 patients because of verapamil-induced hypotension. Mason et al. [45] found verapamil to be effective in preventing VT induction in only 3 of 16 patients with chronic recurrent VT. Buxton et al. [46] also found that intravenous verapamil 10 mg had no effect on nonsustained VT induced by programmed electrical stimulation of the heart in 13 patients. The drug did not prevent or alter the mode of VT induction, nor did it slow the rate of the inducible tachycardia. On the other hand, in 3 of the patients the induced VT accelerated, became more prolonged and deteriorated into ventricular fibrillation in 2. The authors concluded that the mechanism of the tachyarrhythmia was unlikely to be accountable in terms of triggered activity due to afterdepolarizations.

It appears that the type of the VT occurring in patients with organic heart disease is probably not related to slow-channel abnormality. In contrast, a recent report [38] has suggested that there was abolition of PVCs in 7 of 8 patients who exhibited persistent frequent PVCs during the first 48 h of acute myocardial infarction. Favorable responses have been reported also in sustained arrhythmias in this setting. For example, Sclarovsky et al. [47] administered intravenous verapamil (3–5 mg) to 8 patients with multiform accelerated idioventricular rhythm occuring during the first 12 h of acute myocardial infarction. The arrhythmia was abolished in 6, slowed in 1 and was without effect in the remaining patient. Massumi et al. [48] also have reported a high success rate in the use of verapamil in controlling parasystolic VT, raising the possibility of triggered automaticity being due to an abnormality of the slow channel. Successful control of polymorphous VT by verapamil treatment was recently reported by Grenadier et al. [49] in 3 of 4 patients during acute myocardial infarction. These 3 patients had failed on lidocaine, Class I agents, overdrive pacing and repeated cardioversion. However, these are uncontrolled studies which raise the possibility but do not provide conclusive proof that calcium antagonism by verapamil might be effective in controlling ventricular tachyarrhythmias. It should be considered that the apparent salutary effect might result from improved ischemia and myocardial conduction rather than as a consequence of a direct antiarrhythmic effect of the drug in ventricular muscle or the Purkinje fibers. Improvement in ischemia in this situation might be effective in reversing reentry, triggered as well as enhanced automaticity. Moreover, it must be emphasized that in the context of myocardial ischemia, verapamil and other CAs may exert a potent effect in reversing coronary artery spasm vasoconstriction and thus relieve ischemia and ischemia-induced ventricular tachyarrhythmia. Such an effect must be regarded as a "secondary" antiarrhythmic effect which is likely to be most readily apparent in the setting of Prinzmetal variant angina [50].

Exercise-Triggered Ventricular Tachycardia

There have been several recent reports which have suggested that exercise-triggered VT having the morphologic pattern of left bundle branch block and right axis deviation might respond predictably and promptly to intravenous verapamil. Such an arrhythmia may occur in patients without identifiable cardiac disease [51]. Wu et al. [52] described 3 such patients; in none was the tachycardia inducible by programmed electrical stimulation, but could be provoked by isoproterenol infusion. Intravenous verapamil terminated the tachycardia in all 3 and prevented its occurrence during exercise in 2. It is of interest that propranolol and lidocaine also prevented the occurrence of the arrhythmia. Three similar cases have been reported by Sung et al. [44] although in this series intravenous verapamil was not effective in preventing the VT inducibility by isoproterenol but the QRS axis in these patients during the tachycardia was not stated. Thus, it is unclear whether the nature of the arrhythmia was identical with that in the patients reported by Wu et al. [53]. At present, it is not clear whether the VT in question is due to an automatic focus or to triggered automaticity [54]; nor is it known whether the observed response to intravenous verapamil in a small series of patients is a property unique to the drug or a common property of all type I CAs. In this context, it is also of interest that Coumel and Attuel [54] recently reported a group of patients without demonstrable heart disease who experienced syncopal spells due to torsades de pointes in association with a normal QT and an exceptionally short coupling interval of the beat initiating the tachycardia. In the authors' experience, the arrhythmia in these patients selectively responded to intravenous and oral verapamil, whereas they were resistant to other antiarrhythmic agents including amiodarone and β-blocking drugs.

Idiopathic Ventricular Tachycardia with Right Bundle Branch Block and Left Axis Deviation Morphology

It is now increasingly appreciated that in young patients without demonstrable cardiac disease, sustained VT that occurs in a few has the morphologic pattern of right bundle branch block with left axis deviation [55–58]. This arrhythmia appears to constitute a distinct clinical entity; it is usually possible to initiate the tachycardia by both atrial as well as ventricular stimulation.

It is of interest that in most cases the VT in these patients responds to verapamil given intravenously [55, 58], whereas β-blockers or class I agents either fail, or exert only a partial response [55, 58]. The precise mechanism for the arrhythmia is not known; electrophysiologic studies have suggested a reentrant mechanism in the hands of some investigators [55, 58] and triggered automaticity in those of others [42] but the difficulties of such an approach in the clinical laboratory are well known [6].

Torsades de Pointe and Calcium Antagonism

There are emerging data to indicate that the most likely cellular basis for torsades de pointe might be early depolarization [59, 60], which appears to be mediated by calcium fluxes [60]. Preliminary data [61] suggest that specific CAs, at least in the experimental model, abolish or prevent the development of early afterdepolarizations [61] and torsades [62]. The significance of these findings in the clinical syndrome of torsades de pointe needs critical study.

Calcium Antagonists and Sudden Death

An outstanding question that needs to be determined is the potential of CAs to prevent sudden cardiac death in patients with coronary artery disease, especially in survivors of acute myocardial infarction. The role of coronary vasospasm in the genesis of sudden death in patients with ischemic heart disease is still unclear. In the case of β-adrenoceptor-blocking drugs, there is evidence that sudden cardiac death can be reduced significantly by prophylactic treatment in the survivors of infarction. It is, however, not known whether the primary action is on electrical instability induced by myocardial ischemia. CAs are potent antianginal drugs with the additional ability to reverse coronary vasospasm. CAs also have the property of minimizing myocardial ischemic damage following experimental coronary occlusion. One might therefore expect that, as a class, they should exert a salutary effect in the preservation of the ischemic myocardium during chronic prophylactic administration to patients with ischemic heart disease and to alter the incidence of ventricular arrhythmias. Controlled trials are needed to test this theoretical consideration. The results of the initial trials have been equivocal and are not particularly encouraging.

Summary and Conclusions

In recent years calcium channel blockers have emerged as a new class of antiarrhythmic agents for the control of certain supraventricular and ventricular arrhythmias. Electrophysiologically, they are heterogeneous, but their main action is mediated through a depressant effect on the slow-calcium channel in cardiac muscle. In isolated muscle, their actions are modulated by their reflex actions and by their interaction with the autonomic nervous system due to the noncompetitive adrenergic-blocking actions that some of the compounds exhibit. The major agents exerting antiarrhythmic actions are verapamil, diltiazem, gallopamil, tiapamil and bepridil; the dihydropyridines are devoid of electrophysiologic actions in vivo. CAs prolong intranodal conduction time, lengthen the effective and functional refractory periods in the A-V node but exert little or no effect on atrial, ventricular, His-Purkinje or bypass tract conduction or refractoriness (except in the case of bepridil which has additional electrophysiologic properties). These effects form the basis of the clinical antiarrhythmic effects of this class of agents. The most striking action is the predictable and prompt termination of the reentrant supraventricular tachycardia by intravenous verapamil and diltiazem and the slowing of the ventricular response in atrial flutter and fibrillation. These agents may also be of value in the chronic control of ventricular response in atrial flutter and fibrillation; their role in multifocal atrial tachycardia and other ectopic tachycardias are less well defined.

CAs reverse ischemic ventricular arrhythmias due to coronary artery spasm, but exert little or no action in the usual forms of sustained ventricular tachyarrhythmias associated with severe structural heart disease. They are poor suppressants of PVCs. Recent data have established their role in exercise-induced tachycardia occurring in the context of ischemic heart disease; they are also of value in VT occurring in young subjects developing the tachycardia with a right bundle branch block with left axis deviation morphology, an arrhythmia thought to be due to triggered automaticity. Experimental data suggest a potentially significant role of CAs in the treatment of torsades de pointe induced by early afterdepolarizations. The role of CAs in reducing the incidence of sudden death in the survivors of acute myocardial infarction remains unclear.

References

1 Singh, B.N.; Vaughan Williams, E.M.: A fourth class of antiarrhythmic action? Effects of verapamil on ouabain toxicity on atrial and ventricular intracellular potentials and on other features of cardiac function. Cardiovasc Res. *6:* 109–119 (1972).

2 Singh, B.N.; Ellrodt, G.; Peter C.T.: Verapamil: a review of its pharmacological properties and therapeutic use. Drugs *15:* 169–197 (1978).

3 Hauswirth, O.; Singh, B.N.: Ionic mechanisms in heart muscle in relation to the genesis and pharmacologic control of cardiac arrhythmias. Pharmac. Rev. *30:* 5–63 (1979).

4 Zipes, D.P.; Heger, J.J.; Prystowsky, E.N.: Pathophysiology of arrhythmias: clinical electrophysiology. Am. Heart J. *106:* 812–820 (1983).

5 Rosen, M.R.; Redder, R.F.: Does triggered activity have a role in the genesis of cardiac arrhythmias? Ann. intern. Med. *96:* 794–801 (1981).

6 Brugada, P.; Wellens, H.J.J.: The role of triggered activity in clinical ventricular arrhythmias. Pace *7:* 260–271 (1984).

7 Kato, R.; Singh, B.N.: Effects of bepridil on the electrophysiologic properties of isolated canine and rabbit myocardial fibers. Am. Heart J. *109:* 271–285 (1986).

8 Josephson, M.E.; Kastor, J.A.: Supraventricular tachycardia: mechanism and management. Ann. intern. Med. *87:* 869–878 (1977).

9 Heng, M.K.; Singh, B.N.; Roche, A.H.G.; Norris R.M.; Mercer, C.H.: Effects of intravenous verapamil on cardiac arrhythmias and on the electrocardiogram. Am. Heart J. *90:* 487–498 (1975).

10 Singh, B.N.; Nademanee, K.; Baky, S.: Calcium antagonists: clinical uses in treating arrhythmias. Drugs *25:* 125–153 (1983).

11 Schamroth, L.D.; Krikler, M.; Garrett, C.: Immediate effects of intravenous verapamil in cardiac arrhythmias. Br. Heart J. *1:* 660–664 (1972).

12 Sung, R.J.; Elser, B.; Mc Allister, R.G.: Intravenous verapamil for termination of reentrant supraventricular tachycardias. Intracardiac studies correlated with plasma verapamil concentrations. Ann. intern. Med. *93:* 682–689 (1980).

13 Rinkenberger, R.L.; Prystowsky, E.N.; Heger, J.J.; Troup, P.J.; Jackman, W.M.; Zipes, D.P.: Effects of intravenous and chronic oral verapamil administration in patients with supraventricular tachycarrhythmias. Circulation *62:* 996–1010 (1980).

14 Rozansky, J.J.; Zaman, L.; Castellanos, A.: Electrophysiologic effects of diltiazem hydrochloride on supraventricular tachycardia. Am. J. Cardiol. *49:* 621–628 (1982).

15 Bertin, A.; Chaitman, B.R.; Bourassa, M.G.; Breuess, G.; Scholl, J.M.; Bruneau Gague, P.; Chabot, M.: Beneficial effect of intravenous diltiazem in the acute management of supraventricular tachyarrhythmias. Circulation *67:* 88–94 (1983).

16 Hung, J.S.; Yeh, S.J.; Liu, F.C.; Fu, M.; Lee, Y.S.; Wu, D.: Usefulness of intravenous diltiazem in predicting subsequent electrophysiologic and clinical responses to oral diltiazem. Am. J. Cardiol. *54:* 1259–1262 (1984).

17 Yeh, S.H.; Liu, F.C.; Chan, Y.Y.; Hung, J.S.; Wu, D.: Termination of paroxysmal supraventricular tachycardia with single oral dose of diltiazem and propranolol. Circulation *71:* 104–109 (1985).
18 Wu, B.; Hung, J.; Kuo, C.; Hsu, K.; Shieth, W.: Effects of quinidine on AV nodal reentrant paroxysmal tachycardia. Circulation *64:* 169–170 (1981).
19 Swiryn, S.O.; Bauernfeind, R.A.; Wyndham, C.R.C.; Dhingra, R.; Palileo, E.; Strasberg, B.; Rosen, K.M.: Effects of oral disopyramide phosphate on induction of paroxysmal supraventricular tachycardia. Circulation *64:* 170–178 (1981).
20 Sethi, K.K.; Jaishankar, S.; Gupta, M.P.: Salutary effect of intravenous ajumaline in patients with paroxymal supraventricular tachycardia mediated by dual atrioventricular nodal pathways by blockade of retrograde pathways. Circulation *70:* 876–883 (1987).
21 Belhassen, B.; Pelleg, M.: Acute management of paroxysmal supraventricular tachycardia: verapamil, adenosine triphosphate or adenosine? Am. J. Cardiol. *54:* 225–227 (1984).
22 Belhassen, B.; Glick, A.; Laniado, S.: Comparative clinical and electrophysiologic effects of adenosine triphosphate and verapamil on paroxysmal reciprocating junctional tachycardia. Circulation *77:* 795–805 (1988).
23 Opie, L.H.; Coetzee, W.A.; Dennis, C.; Thandroyen, F.T.: A potential role of calcium ions in early ischemic and reperfusion arrhythmias. Ann. N.Y. Acad. Sci. *522:* 464–477 (1988).
24 Aronow, W.S.; Plasencia, G.; Wong, R.: Effect of verapamil versus placebo in PAT and MAT (paroxysmal atrial tachycardia and multifocal atrial tachycardia). Curr. ther. Res. *27:* 823–826 (1980).
25 Levine, H.J.H.; Michael, J.R.; Guarnieri, T.: Treatment of multifocal atrial tachycardia with verapamil. New Engl. J. Med. *312:* 21–25 (1985).
26 Mauritson, D.R.; Winniford, M.D.; Walker, W.S.; Rude, R.E.; Cary, J.R.; Hills, L.D.: Oral verapamil for paroxysmal supraventricular tachycardia. A long-term, double-blind randomized trial. Ann. intern. Med. *96:* 409–412 (1982).
27 Sakurai, M.; Yasuda, I.T.; Kato, N.; Nomura, A.; Fujita, M.; Nishino, T.; Fujita, K.; Kolke, Y.; Saito, H.: Acute and chronic effects of verapamil in patients with paroxysmal supraventricular tachycardia. Am. Heart J. *105:* 619–628 (1982).
28 Pritchett, E.L.C.; Hammill, C.; Reiter, M.J.; Lee, K.L.; McCarthy, E.A.; Zimmerman, J.M.; Shand, D.G.: Life-table method for evaluating antiarrhythmic drug efficacy in patients with paroxysmal atrial tachycardia. Am. J. Cardiol. *52:* 1007–1014 (1983).
29 Winniford, M.D.; Fulton, M.D.; Hillis, D.L.: Long-term therapy of paroxysmal supraventricular tachycardia: a randomized double-blind comparison of digoxin, propranolol and verapamil. Am. J. Cardiol. *54:* 01138–1139 (1984).
30 Tonkin, A.M.; Aylward, P.E.; Joel, S.E.: Verapamil on prophylaxis of paroxysmal atrioventricular nodal reentrant tachycardia. J. Cardiovasc. Pharm. *2:* 473–480 (1980).
31 Klein, G.J.; Gulamhusein, S.; Carruthers, S.G.; Donner, A.P.; Ko, P.T.: Comparison of the electrophysiologic effects of intravenous and oral verapamil in patients with paroxysmal supraventricular tachycardia. Am. J. Cardiol. *49:* 117–124 (1982).
32 Storstein, O.; Landmark, K.H.: Verapamil in the treatment of atrial tachycardia with block. Acta med. scand. *198:* 483–488 (1975).
33 Khalsa, A.; Olssen, S.B.: Verapamil-induced ventricular regularity in atrial fibrillation. Acta med. scand. *205:* 509–515 (1979).

34 Klein, H.O.; Kaplinsky, E.: Digitalis and verapamil natrial fibrillation and flutter. Is verapamil now the preferred agent? Drugs *31:* 185–197 (1986).

35 Roth, A.; Harrison, E.; Mitani, G.; Jordan, C.; Rahimtoola, S.H.; Elkayam, U.: Efficacy and safety of medium- and high-dose diltiazem alone and in combination with digoxin for the control of heart rate at rest and during exercise in patients with chronic atrial fibrillation. Circulation *73:* 316–325 (1986).

36 Basu, B.; Cherian, G.; Abraham, K.A.: Oral verapamil as maintenance therapy after cardioversion for atrial fibrillation. Ind. J. Chest Dis. Allied Sci. *19:* 170–176 (1977).

37 Rasmussen, K.; Wang, H.; Faisa, D.: Comparative efficacy of quinidine and verapamil in the maintenance of sinus rhythm after DC conversion of atrial fibrillation. Acta med. scand., suppl. *645:* pp 23–28 (1981).

38 Belhassen, B.; Horowitz, L.N.: Role of intravenous verapamil for ventricular tachycardia. Am. J. Cardiol. *54:* 1141–1133 (1984).

39 Krikler, D.M.; Goodwin, J.F.: Arrhythmia as a cause of sudden death in hypertrophic cardiomyopathy. Lancet *ii:* 937–938 (1976).

40 McKenna, W.J.; Harris, L.; Perez, G.: Hypertrophic cardiomyopathy: Comparison of verapamil and amiodarone in the treatment of arrhythmia. Br. Heart J. *45:* 354–360 (1980).

41 Kaumann, A.J.; Aramendia, P.: Prevention of ventricular fibrillation induced by coronary ligation. J. Pharmac. exp. Ther. *164:* 326–332 (1968).

42 Wellens, H.J.J.; Bar, F.W.; Lie, K.I.; Duren, D.R.; Dohmen, H.J.: Effects of procainamide, propranolol and verapamil on mechanism of tachycardia in patients with chronic recurrent ventricular tachycardia. Am. J. Cardiol. *40:* 579–585 (1977).

43 Wellens, H.J.J.; Farre, J.; Bar, W.B.: The role of the slow inward current in the genesis of ventricular tachyarrhythmias in man; in Zipes, Bailey, Elharrar, The slow inward current and cardiac arrhythmias, pp. 507–514 (Nijhoff, Boston, 1980).

44 Sung, R.J.; Shapiro, W.A.; Shein, E.W.; Morady, F.; Davis, J.: Effects of verapamil on ventricular tachycardias possibly caused by reentry, automaticity, and triggered activity. J. clin. Invest. *72:* 350–360 (1983).

45 Mason, J.W.; Swerdlow, C.D.; Mitchell, L.B.: Efficacy of verapamil in chronic, recurrent ventricular tachycardia. Am. J. Cardiol. *51:* 1614–1617 (1983).

46 Buxton, A.E.; Waxman, H.L.; Marchlinski, F.E.; Josepson, M.E.: Electropharmacology of non-sustained ventricular tachycardia: effects of class I antiarrhythmic agents verapamil and propranolol. Am. J. Cardiol. *53:* 738–744 (1984).

47 Sclarovsky, S.; Strasberg, B.; Fuchs, J.; Lewin, R.F.; Arditi, A.; Klainman, E.; Kracoff, O.H.; Agmon, J.: Multiform accelerated idioventricular rhythm in acute myocardial infarction: electrocardiographic characteristics and response to verapamil. Am. J. Cardiol. *52:* 43–47 (1983).

48 Massumi, R.A.; Martin, J.; Udhohi, V.N.: Verapamil in parasystolic ventricular tachycardia and other wide QRS rhythms. Am. Heart J. *11:* 400–405 (1986).

49 Grenadier, E.; Alpan, G.; Maor, N.; Keidar, S.; Birnenboim, C.; Marguiles, T.; Palant, A.: Polymorphous ventricular tachycardia in acute myocardial infarction. Am. J. Cardiol. *53:* 1280–1283 (1984).

50 Kimura, E.; Tanaka, K.; Mizuno, K.; Handa, Y.; Hashimoto, H.: Suppression of repeatedly occurring ventricular fibrillation with nifedipine in variant form of angina pectoris. Jap. Heart J. *18:* 736–742 (1977).

51 Palileo, E.V.; Ashley, W.W.; Swiryn, S.; Bauernfeind, R.A.; Strasberg, B.; Petrogoulos, A.T.; Rosen, K.M.: Exercise provocable right ventricular outflow tract tachycardia. Am. Heart J. *104:* 185–193 (1982).
52 Wu, D.; Kou, H.C.; Hung, J.S.: Exercise-triggered paroxysmal ventricular tachycardia. A repetitive rhythmic activity possibly related to afterdepolarization. Ann. intern. Med. *95:* 410–414 (1981).
53 Wu, D.: What is the value of calcium antagonists for ventricular tachycardia? Int. J. Cardiol. *5:* 543–547 (1984).
54 Coumel, P.; Attuel, P.: Which arrhythmias are specifically susceptible to calcium antagonists; in Rosenbaum, Elizari, Frontiers of cardiac electrophysiology, pp. 341–348 (Nijhoff, Boston, 1983).
55 Belhassen, B.; Rotmensch, H.H.; Laniado, S.: Response of recurrent sustained ventricular tachycardia to verapamil. Br. Heart J. *46:* 679–682 (1981).
56 Lin, F.C.; Finley, D.; Rahimtoola, S.H.; Wu, D.: Idiopathic paroxysmal ventricular tachycardia with a QRS pattern of right bundle branch block and left axis deviation: a unique clinical entity with specific properties. Am. J. Cardiol. *52:* 96–100 (1983).
57 German, L.D.; Packer, D.L.; Brady, G.H.; Gallagher, J.J.: Ventricular tachycardia induced by atrial stimulation in patients without symptomatic cardiac disease. Am. J. Cardiol. *52:* 1202–1207 (1983).
58 Belhassen, B.; Shapira, I.; Padleg, A.; Coperman, I.; Kauli, N.; Laniado, S.: Idiopathic recurrent sustained ventricular tachycardia responsive to verapamil: an ECG-electrophysiologic entity. Am. Heart. J. *108:* 1034–1036 (1984).
59 El-Sherif, N.; Zeiler, R.; Craelius, W.; Henkin, R.; Gough, W.: Bradycardia-dependent QTc prolongation and torsades de pointes due to early afterdepolarization. Circulation *76:* suppl. IV, pp. IV–429 (1987).
60 January, C.T.; Riddle, J.M.: Early afterdepolarizations and Ca^{++}. Circulation *76:* suppl. IV, pp. IV–472 (1987).
61 Takanaka, C.; Singh, B.N.: Barium-induced non-driven action potentials as a model of triggered automaticity and early afterdepolarizations: differing effects of amiodarone and class I agents and significance of calcium-channel blockade. Submitted for publication (1988).
62 Aliot, E.; Szabo, B.; Sweidan, R.; Lazzara, R.: Prevention of torsades de pointes with calcium channel blockade in an animal model. J. Am. Coll. Cardiol. *5:* 492 (1985).

Bramah N. Singh, MD, Wadsworth VA Medical Center, Cardiology Section (691/111E), Wilshire and Sawtelle Blvds., Los Angeles, CA 90073 (USA)

Van Zwieten PA (ed): Clinical Aspects of Calcium Entry Blockers.
Prog Basic Clin Pharmacol. Basel, Karger, 1989, vol 2, pp 88–102

Treatment of Myocardial Ischemia with Calcium Entry Blockers

W. H. van Gilst, K. I. Lie

Department of Cardiology, Groningen, The Netherlands

Introduction

Calcium plays a vital role in many biologic processes including the cardiovascular system, where calcium couples electrical activation to muscular contraction, both in the myocardium and vascular smooth muscle. Thus, calcium plays a key role in cardiac contraction and vasoconstriction. The development of calcium entry blockers (CEBs) has given the clinician several important new therapeutic agents for use in a variety of disease states. Especially, in the treatment of ischemic heart disease this new class of agents has been most useful. The purpose of this review is twofold: first, to describe the causes leading to the different syndromes in the broad spectrum of ischemic heart disease, and, second, to identify the possible role of CEBs in the treatment of patients with these ischemic syndromes.

Pathogenic Classification of Ischemic Heart Disease

Ischemic heart disease consists of a broad spectrum of syndromes, all with a different underlying pathogenic mechanism. Therefore, for the treatment of ischemic heart disease, the cause of myocardial ischemia must first be identified.

Three main causes have been implicated in the development of ischemic heart disease. The first is a fixed atherosclerotic stenosis, a permanent narrowing of the coronary arteries, with pain occurring only upon exertion. The second, the focus of much recent interest, is spasm, in which dynamic changes in vessel diameter cause transient arterial obstruction. Third, excessive platelet aggregation can lead to partial, and occasionally total, coronary vessel blockage and myocardial infarction, often coincident with excessive thromboxane A_2 release and spasm. All of these causes interact and produce together the spectrum of ischemic heart disease. This interaction is a dynamic process and may change in time within the individual patient. Therefore, each patient may present with different ischemic syndromes.

Originally, it was thought that stable angina or exertional angina is exclusively caused by a fixed atherosclerotic lesion. However, during recent years it has been suggested that coronary artery spasm is far commoner in patients with stable angina [Conti, 1985; Maseri et al., 1977, 1985]. Patients suffering from this form of angina have a recognizable threshold of exertion that they can never exceed without developing angina. In addition, they have episodes of angina at rest or at levels of exertion that are quite well tolerated. The term 'mixed angina' was proposed for this form [Maseri, 1980] and mainly two of the above-mentioned elements are thought to determine this syndrome.

The first is the critical coronary obstruction that sets the limit of maximal residual coronary flow reserve, i. e., the level of effort or of oxygen demand that the patient cannot exceed without developing ischemia. This maximal residual coronary flow reserve results from the balance between the severity of coronary obstruction and collateral blood supply. The second is the occurrence of coronary spasm which can suddenly and transiently reduce coronary flow to a variable extent. When flow is reduced below the resting needs, spontaneous ischemia at rest occurs. The mechanisms that may be responsible for the transient impairment of coronary blood flow caused by increased vascular tone include supersensitivity of a segment of a large epicardial coronary artery. This causes a total or subtotal occlusion in response to stimuli that will normally change vascular diameter only to a modest degree.

In these relative stable forms of ischemic heart disease the eventual syndrome, i.e., stable, mixed or variant angina, is determined by the balance between the presence of these factors.

However, sudden changes in the balance of atherosclerosis and spasm on the one hand and platelet function on the other may cause ischemic syndromes which are more unstable in character. As recently reviewed by Fuster et al. [1985], rupture, cracking, or ulceration of atherosclerotic plaques, or a combination, is a common autopsy finding in patients with ischemic heart disease [Falk, 1983, 1985; Davies and Thomas, 1984, 1985]. Nearly half of the ruptured atherosclerotic plaques have associated mural or occlusive thrombi. Conversely, and most important, many coronary artery thrombi are anchored in cracks or in ruptured atherosclerotic plaques; the part of the thrombus directly overlying the fissure is predominantly platelet related. Recent approaches and technical improvements, such as serial coronary arteriography [Mois et al., 1983], reperfusion by thrombolysis [Rentrop, 1985; Mandelkorn et al., 1983], operation during acute coronary artery syndromes [DeWood et al., 1980], postmortem coronary arteriography [Levin and Fallon, 1982] and histopathologic studies have emphasized the importance of this process of plaque rupture and thrombosis. Such acute or subacute change in the balance of the underlying factors responsible for myocardial ischemia appears to be the most frequent cause of acute ischemic syndromes such as unstable angina [Holmes et al., 1981; Ambrose et al., 1985; Sherman et al., 1985] and myocardial infarction.

Since the efficacy of CEBs in the treatment of ischemic heart disease is largely dependent on the underlying pathology, the results of the major clinical trials will be discussed in the light of the above-mentioned characterization and subclassification of ischemic syndromes.

Results of Controlled Trials with Calcium Entry Blockers in Relatively 'Stable' Ischemic Syndromes, i.e., Stable and Mixed Angina

In stable exertional angina, fixed coronary obstruction due to atherosclerosis does not usually compromise coronary flow at rest. If, however, the requirements for coronary flow exceed a maximal amount that gets through the stenosis, ischemia and subsequent angina supervene. In addition to this fixed atherosclerotic stenosis, nonphysiologic changes in coronary vascular tone may cause dynamic limitation of coronary flow superimposed on the existing maximal flow reserve. This last form is often called mixed angina. The most important determinants of myocardial oxygen demand are heart rate, myocardial stress and inotropic state of the heart, in which myocardial stress is determined by afterload, size and shape of the heart; therefore stable effort angina usually occurs on physical exertion, emotion and exposure to cold. The basic principles of medical treatment are based on reduction of heart rate, pre- and/or afterload.

CEBs have been studied extensively in well-designed randomized, placebo-controlled trials in stable effort angina. In these studies verapamil, nifedipine and diltiazem were all effective as monotherapy in reducing spontaneous anginal attacks and exercise-induced ischemia during ergometry (table I).

Since all CEBs reduce afterload, whereas diltiazem and verapamil have negative chronotropic effects, it is conceivable that the beneficial effects of these drugs in stable angina are based on reduction of afterload and/or heart rate. In fact, clinical studies show some reduction of heart rate, blood pressure and rate pressure product at rest, but no significant reduction of rate pressure product at peak exercise, except for verapamil. Although a direct negative inotropic effect of a CEB may be responsible for the reduction of myocardial oxygen demand, in vivo studies [Belanger et al., 1986] show that these negative inotropic effects are counteracted by afterload reduction. Even in patients with depressed myocardial function, diltiazem and especially nifedipine may result in a rise in cardiac output due to predominating peripheral vasodilatation. Alternative antianginal mechanisms of CEBs relate to improvement of coronary flow in ischemic zones [Bache and Dymek, 1982] and/or reduction of ischemic injury currents by improving intracellular calcium homeostasis [Clusin, 1982], which is only proven in the experimental setting. The observation that total coronary blood flow and left ventricular ejection fraction increase following diltiazem or nifedipine treatment [Belanger et al., 1986] support this hypothesis.

CEBs as a monotherapy have been compared to β-blockers. In most studies, it appeared that moderate to high doses of CEBs are as effective as β-blockers in improving anginal complaints and exercise tolerance [Hung et al., 1983; Frishman et al., 1982; Higginbotham et al., 1986]. The incidence of side effects seem to be acceptable in patients treated with CEBs but is especially low in those treated with diltiazem [Nayler and Horowitz, 1983].

CEBs combined with β-blockers in stable angina have potential pharmacodynamic antianginal benefits because the reflex coronary and systemic vasoconstriction mediated by β-blockade is neutralized by the CEB. In fact, most clinical studies comparing the combination of CEBs with β-blockers versus CEBs or β-blockers as monotherapy do show additional substantial improvement of symptoms and exercise tolerance [Boden et al., 1985; Uusitalo et al., 1986; Winniford et al., 1985]. In these studies, addition of CEBs to β-blockade resulted in a further substantial decrease of rate pressure product at peak exercise, suggesting that the

Table I. Effects of CEBs on exercise capacity in stable effort angina in randomized placebo-controlled trials

Drug and dose (mg)	Improvement vs. placebo	Authors
Diltiazem 120–240	$p<0.001$ (240 mg)	Hossack et al., 1981
Diltiazem 120–240	$p<0.01$ (240 mg)	Low et al., 1981
Diltiazem 240	$p<0.05$	Pool and Seagren, 1982
Diltiazem 120–240	$p<0.002$ (240 mg)	Hossack et al., 1982
Diltiazem 120–240	$p<0.01$	Strauss et al., 1982
Diltiazem 120	$p<0.01$	Wagniart et al., 1982
Diltiazem 360	$p<0.01$	Hung et al., 1983
Nifedipine 30	$p<0.005$	Moskowitz et al., 1979
Nifedipine 30	$p<0.01$	Ekelund and Oro, 1979
Nifedipine 30	$p<0.005$	Jaiprakash et al., 1980
Nifedipine 30–60	$p<0.002$	Mueller and Chahine, 1981
Nifedipine 30–60	$p<0.05$	Sherman and Chang, 1983
Verapamil 240	$p<0.05$	Fagher et al., 1977
Verapamil 360	$p<0.05$	Simoons et al., 1980
Verapamil 360	$p<0.001$	Dawson et al., 1980
Verapamil 240–480	$p<0.001$	Pine et al., 1982
Verapamil 360	$p<0.00$	Subramian et al., 1982
Verapamil 240–480	$p<0.001$	Frishman et al., 1982

beneficial effects of calcium entry blockade in these patients are mainly due to a decrease of myocardial oxygen demand rather than an improvement of myocardial oxygen supply.

In one study [Johnston et al., 1985] comparison of a β-blocker with either verapamil, diltiazem or nifedipine revealed the lowest incidence of side effects with diltiazem, with equipotent antianginal effects. The combination of verapamil and β-blockers is associated with frequent lengthening of the PR interval and occasional symptomatic bradycardia [Belanger et al., 1986; Johnston et al., 1985], whereas the nifedipine-β-blocker combination was associated with an increase in ejection fraction and cardiac index [Johnston et al., 1985]. This suggests that combination of verapamil and β-blockers should be given with caution in patients with bradycardia or A-V conduction disturbances, whereas a combination of nifedipine and β-blockers has to be considered in those with depressed myocardial function.

In summary, the efficacy of CEBs in stable effort angina seems to be well established for both monotherapy and combination therapy with β-blockers. The indication for monotherapy is evident when contraindications or serious side effects due to β-blockers are present. Approximately 20% of the population does tolerate β-blockers because of these reasons. Under these circumstances, diltiazem and verapamil are preferred over nifedipine because of their negative chronotropic effects. Only if heart failure or severe bradycardia constitute the contraindications for β-blockade, is nifedipine preferred, because of its powerful unloading effects, which often lead to reflex tachycardia. Combined therapy with β-blockers is indicated when monotherapy with β-blockers fails. Combined therapy with nifedipine or diltiazem is preferred over verapamil because of the marked negative inotropic and dromotropic effects of the latter. CEBs may be combined with nitrates, when β-blockers are contraindicated and monotherapy with either CEBs or nitrates failed. Under these conditions, verapamil or diltiazem is preferred in order to avoid excessive reflex tachycardia.

Results of Controlled Trials with Calcium Entry Blockers in Acute Unstable Ischemic Syndromes, i.e., Unstable Angina and Myocardial Infarction

Unstable Angina

In recent years there have been many studies on the natural history and treatment of patients with angina at rest or 'unstable angina', often characterized as a clinical syndrome between exertional angina or 'stable angina pectoris' and acute myocardial infarction. However, in reviewing published reports there appear to be major drawbacks that make clinical application of these studies difficult. First, the studies are not uniform with respect to the definition of unstable angina, as indicated by their inclusion criteria, resulting in analyses of heterogeneous patient populations. With the currently available diagnostic methods it is impossible to differentiate unstable angina from evolving myocardial infarction at admission of a patient to a coronary care unit. Many of the patients with suspected unstable angina have already sustained a myocardial infarction or are in the process of doing so. Differences in inclusion criteria for the duration of chest pain and electrocardiographic evidence of acute infarction will lead to the above-mentioned heterogeneity in patient population. Second, there is a sizable cross-over rate from medical to surgical therapy in many trials. This variability affects the interpretation of these trials and makes comparison of similar trials difficult.

Table II summarizes the results of four major randomized double-blind studies of CEBs in patients with unstable angina. In all these studies, nifedipine was used. Data from large clinical studies using other CEBs are not available at present.

As can been seen from table II, there are substantial differences in the rates of progression to myocardial infarction. Gerstenblith et al. [1982] showed that nifedipine reduces the recurrence rate of all ischemic events in the first 3 months after starting the treatment. However, in this study nifedipine or placebo was added to a standard regimen of propranolol and long-acting nitrates. Muller et al. [1984a, b] studied two groups of patients with an acute ischemic syndrome, one defined as patients with 'threatened myocardial infarction' [Muller et al., 1984a] and the other as patients with 'unstable angina' [Muller et al., 1984b]. The second group of patients appeared to be far more healthy (see table II). However, nifedipine had no effect in either group when compared to placebo [Muller et al., 1984a] or propranolol plus long-acting nitrates [Muller et al., 1984b]. In the HINT [1986] study, aspects of the three above-mentioned studies were combined. Nifedipine, metoprolol or the combination of both drugs were studied, placebo controlled, in a large group of patients (515) at intermediate risk for developing myocardial infarction (25% during the first week after start of treatment). There was no difference in the percentage of patients developing myocardial infarction or recurrent myocardial ischemia during the first 48 h after starting the treatment. In addition, this study confirmed that in patients already treated with β-blockers, the addition of a CEB might be advantageous.

In summary, if CEBs were to be of use in unstable angina then it would be most appropriate to use them either in patients already on β-blockers or in a subset of patients were spasm plays a more predominant role than platelet activation. A difficulty with the last proposition is that it is impossible at present to distinguish the various subsets of patients with an acute ischemic syndrome. Furthermore, the studies with nifedipine suggest that the latter subset of patients is relatively small in number. A greater understanding of the pathophysiology of unstable angina and early infarction, and improved techniques for subsetting patients are essential if such an approach is to be used.

Acute Myocardial Infarction

Numerous studies, both experimental and clinical, have been conducted over the past 10 years on the ability of CEBs to reduce ischemic

Table II. Differences in the rates of progression to myocardial infarction (MI) according to drug treatment

Experimental treatment	Results	Authors
β-Blocker + nifedipine vs. β-blocker + placebo	Reduced recurrent ischemia during 3 months compared with placebo	Gerstenblith et al., 1982
Nifedipine vs. conventional therapy	Progression to MI 14%/14 days; no difference with conventional therapy	Muller et al., 1984b
Nifedipine vs. placebo	Progression to MI 75%/24 h; no difference with placebo	Muller et al., 1984a
Nifedipine/β-blocker/nifedipine + β-blocker vs. matching placebo	Progression to MI 25%/7 days; no effect for nifedipine alone	HINT Report, 1986

myocardial necrosis [for reviews, see Moss, 1986; Kloner and Braunwald, 1987]. The rationale for using CEBs to reduce myocardial infarct size includes their demonstrated anti-ischemic properties, the possible role of calcium overload in enhancing myofibrillar necrosis in acute myocardial infarction, and their ability to relieve coronary vasospasm and constriction occurring after myocardial infarction. In many ways, the rationale is comparable to the considerations which dictated the design and execution of the postinfarction β-blockade studies during the 1970s. However, the results of the few available clinical trials with CEBs have been disappointing thus far. In this section, we will evaluate the major reported clinical studies that have examined the effects of CEBs in the setting of acute regional ischemia and infarction.

Table III summarizes the major clinical studies examining the effect of CEBs in patients with myocardial infarction. There have been two major studies that have examined the effect of verapamil on myocardial infarct size [Thuesen et al., 1984; Bussmann et al., 1984] and one large multicenter study with mortality as primary endpoint [The Danish Study Group, 1984; Danish Multicenter Study Group, 1984].

Thuesen et al. [1984] administered verapamil or placebo to patients with acute myocardial infarction at a median of 4 h after the onset of chest pain. Infarct size was measured by cumulative CK values. Verapamil did

Table III. Clinical studies of the effect of calcium entry blockers in myocardial infarction.

Drug	Number of Patients	Endpoint	Effect of therapy	Start therapy after onset of symptoms, h	Authors
Verapamil	100	infarct size	no effect	4	Thuesen et al., 1983
Verapamil	29	infarct size	reduction	8	Bussman et al., 1984
Verapamil	1,436	mortality	no effect	–	Danish Study group 1984
Nifedipine	171	infarct size	no effect	4.6	Muller et al., 1984b
Nifedipine	227	infarct size	no effect	5.5	Sirnes et al., 1984
Nifedipine	4,491	mortality	no effect	–	Wilcox et al., 1986
Diltiazem	576	reinfarction	reduction	24–72	Gibson et al., 1986

not reduce infarct size in this study and the investigators postulated that the negative result might be secondary to the 4-hour delay in administration of the drug. However, Bussmann et al. [1984] randomized patients to verapamil or control groups starting at an average of 8 h after onset of chest pain. Peak CK and CK-MB activity derived from time-activity curves were lower in the verapamil group than in the control group. Thus, in contrast to the theory of Thuesen et al. [1984], this clinical study suggests a reduction of infarct size with verapamil given at a later stage after the onset of symptoms.

In the Danish Multicenter Study [1984], the primary endpoint was not infarct size but mortality. In this study, 717 patients were randomized to receive verapamil on admission followed by therapy for 6 months, while 719 patients received placebo. After 6 months, 12.8% of patients in the verapamil group and 13.9% of the patients in the placebo group had died (difference not significant).

There have been two major studies to determine if nifedipine was capable of decreasing myocardial infarct size in man [Muller et al., 1984b; Sirnes et al., 1984] and one large multicenter study with mortality as primary endpoint [Wilcox et al., 1986]. Muller et al. [1984b] randomized a total of 171 patients – 105 with threatened myocardial infarction and 66 with acute myocardial infarction – to receive either nifedipine (20 mg each 4 h) or placebo, along with standard care. Investigational therapy was initiated at an average of 4.6 h after the onset of chest pain. Nifedipine did not affect the progression to acute infarction among the patients with threatened infarction nor did it alter infarct size, as measured by CK-MB time-activity curves in patients with acute myocardial infarction or in those with threatened infarction who progressed to actual infarction. There was, in fact, a tendency to a higher mortality rate in the nifedipine-treated patients 2 weeks after entry (7.9% vs. 0%), but this difference did not persist at six months (10.1% vs. 8.5%).

The Norwegian Nifedipine Multicenter Trial showed similar results [Sirnes et al., 1984]. Patients with acute myocardial infarction were randomized within 12 h of chest pain to receive either nifedipine (n=112) or placebo (n=115). Nifedipine was administered at an average of 5.5 h after chest pain. In this study nifedipine failed to reduce infarct size, also estimated from CK-MB curves.

In the more recent TRENT study, 4,491 patients with acute myocardial infarction were randomized within 24 h to either nifedipine (10 mg grid; n=2,240) or placebo (n=2,251) [Wilcox et al., 1986]. The primary end-point, the 1-month mortality rate, was not affected by nifedipine treatment (6.3% in the placebo-treated group versus 6.7% in the nifedipine-treated group).

In order to detect a small reduction in mortality in a group of patients with a 10% mortality several thousands of patients need to be included in a study. It is possible, therefore, that these studies have failed to demonstrate a minor benefit that CEBs might have. However, in a recent analysis of pooled data from nine major randomized controlled trials using the technique of retrospective stratification a 6% excess in mortality by treatment was shown [Yusuf and Furberg, 1987]. Therefore, the clinical role of CEBs in the acute phase of myocardial infarction is, on the basis of these data, at best uncertain.

All the studies discussed above were started in the acute phase of myocardial infarction. Recently, positive results of a large multicenter study with diltiazem therapy, initiated in the subacute phase (24–72 h after the onset of symptoms), in patients with non-Q wave infarction were published [Gibson et al., 1986]. The primary goal of this study was to reduce myocardial reinfarction (mortality was about 3% in this patient group) during the 2 weeks of active therapy. The beneficial results on reinfarction observed in this study (a reduction from 9.3 to 5.2%) may have important implications. Non-Q wave myocardial infarction is likely to increase because of the widespread use of thrombolytic therapy which, if successful, would limit the extent of myocardial infarction, resulting in a larger number of patients with non-Q wave infarction [Rentrop et al., 1984; Ganz et al., 1984; DeWood et al., 1980]. Thus, this study not only shows that diltiazem is effective as routine prophylaxis in the routine management of patients recovering from acute non-Q wave myocardial infarction, but it also shows that diltiazem may be effective as potential prophylaxis to be given in conjunction with thrombolytic therapy.

Conclusions

Since the recognition of the concept of calcium entry blockade many experimental and clinical studies have been performed to investigate the potential use of CEBs in ischemic heart disease. Their beneficial role in stable angina and angina of vasospastic origin is well established at present. However, the promising results of CEBs in the experimental setting of acute myocardial infarction or unstable angina are, so far, not reproduced in the clinical situation. This may be explained by an incomplete understanding of the pathophysiology of these ischemic syndromes. Furthermore, not yet fully understood differences between the mechanisms of action of the various CEBs may play a role.

References

Ambrose, J.A.; Winters, S.L.; Stern, A.; Eng, A.; Teichholz, L.E.; Gorlin, R.; Fuster, V.: Angiographic morphology and the pathogenesis of unstable angina pectoris. JACC *5:* 609–616 (1985).

Bache, R.J.; Dymek, D.J.: Effect of diltiazem on myocardial blood flow. Circulation *65:* 1–19 (1982).

Belanger, L.G.; Charbonneau, M.; Lavallee, J.P.; et al.: Treatment of stable angina pectoris with calcium antagonists alone or combined with β-blocking agents. A review of the literature. Can. J. Cardiol. *2:* 212–217 (1986).

Boden, W.E.; Bough, E.W.; Reichman, M.J.: Beneficial effects of high dose diltiazem in patients with persistent effort angina on β-blockers and nitrates. A randomized double-blind placebo-controlled cross-over study. Circulation *71:* 1197–1205 (1985).

Bussmann, W.-D.; Seher, W.O.; Gruengras, M.: Reduction of creatine kinase and creatine kinase-MB indexes of infarct size by intravenous verapamil. Am. J. Cardiol. *54:* 1224–1230 (1983).

Conti, C.R.: Large vessel coronary vasospasm: diagnosis, natural history and treatment. Am. J. Cardiol. *55:* 41B–49B (1985).

Clusin, W.T.: Metabolic inhibitors induced Ca-dependent inward current in cardiac cells. Circulation *66:* suppl., pp. II–15 (1982).

Danish Multicenter Study Group on Verapamil in Myocardial Infarction: Verapamil in acute myocardial infarction. Am. J. Cardiol. *54:* 24E–28E (1984).

Davies, M.J.; Thomas, A.C.: Thrombosis and acute coronary-artery lesions in sudden cardiac ischemiac death. New Engl. J. Med. *310:* 1137–1140 (1984).

Davies, M.J.; Thomas, A.C.: Plaque fissuring – the cause of acute myocardial infarction, sudden ischaemic death, and crescendo angina. Br. Heart J. *53:* 363–373 (1985).

Dawson, J.R.; Whitaker, N.H.; Sutoon, S.C.: Calcium antagonist drugs in chronic stable angina. Comparison of verapamil and nifedipine. Br. Heart J. *46:* 508–512 (1980).

DeWood, M.A.; Spors, J.; Notske, R.; Mouser, L.T.; Burroughs, R.; Golden, M.S.; Lang, H.T.: Prevalence of total coronary occlusion during the early hours, of transmural myocardial infarction. New Engl. J. Med. *303:* 897–902 (1980).

Ekelund, L.G.; Oro, L.: Antianginal efficacy of nifedipine with and without a β-blocker studied with exercise test. Clin. Cardiol. *2:* 203–211 (1979).

Fagher, B.; Svenson, S.; Persson, S.: Double-blind comparison of verapamil and practolol in the treatment of angina pectoris. Post-grad. med. J. *53:* 61–65 (1977).

Falk, E.: Plaque rupture with severe pre-existing stenosis precipitating coronary thrombosis: characteristics of coronary atherosclerotic plaques underlying fatal occlusive thrombi. Br. Heart J. *50:* 127–134 (1983).

Falk, E.: Unstable angina with fatal outcome: dynamic coronary thrombosis leading to infarction and/or sudden death. Circulation *71:* 699–708 (1985).

Farhi, J.-I.; Cohen, M.; Fuster, V.: The broad spectrum of unstable angina pectoris and its implications for future controlled trials. Am. J. Cardiol. *58:* 547–550 (1986).

Frishman, W.H.; Klein, N.A.; Strom, J.A.: Superiority of verapamil in the treatment of angina pectoris: a double-blind randomized crossover trial. Circulation *65:* suppl. 1, pp.51–59 (1982).

Fuster, V.; Steele, P.M.; Chesebro, J.H.: Role of platelets and thrombosis in coronary artherosclerotic disease and sudden death. JACC *5:* 175B (1985).

Ganz, W.; Geft, I.; Shah, P.; Lew, A.S.; Rodriguez, L.; Weiss, T.; Maddahi, J.; Berman, D.S.; Charuzi, Y.; Swan, H.J.C.: Intravenous streptokinase in evolving acute myocardial infarction. Am. J. Cardiol. *53:* 1209–1216 (1984).

Gerstenblith, G.; Ouyang, P.; Achuff, S.C.; et al.: Nifedipine in unstable angina. A double-blind randomised trial. New Engl. J. Med. *306:* 885–889 (1982).

Gibson, R.S.; Boden, W.E.; Theroux, P.; Strauss, H.D.; Pratt C.M.; Gheorghiade, M.; Capone, R.J.; Crawford, M.H.; Schlant, R.C.; Kleiger, R.E.; Young, P.M.; Schechtmann, K.; Perryman, M.B.; Roberts, R.: Diltiazem and reinfarction in patients with non-Q wave myocardial infarction. Results of a double-blind, randomized, multicenter trial. New Engl. J. Med. *315:* 423–429 (1986).

Gottlieb, S.O.; Weisfeld, M.L.; Ouyand, P.; Achuff, S.C.; Baughman, K.L.; Traill, T.A.; Brinker, J.A.; Shapiro, E.P.; Chandra, N.C.; Mellits, E.D.; Townsend, R.N.; Gerstenblith, G.: Effect of the addition of propranolol to therapy with nifedipine for unstable angina pectoris: a randomized, double-blind, placebo-controlled trial. Circulation *73:* 331–337 (1986).

Higginbotham, M.B.; Morris, K.G.; Coleman, R.E.: Comparison of nifedipine alone with propranolol alone for stable angina pectoris including hemodynamics. Am. J. Cardiol. *57:* 1022–1028 (1986).

HINT Research Group: Early treatment of unstable angina in the coronary care unit: a randomised, double-blind, placebo controlled comparison of recurrent ischaemia in patients treated with nifedipine ov metoprolol or both. Br. Heart J. *56:* 400–413 (1986).

Holmes, D.R.; Hartzler, G.O.; Smith, H.C.; Fuster, V.: Coronary artery thrombosis in patients with unstable angina. Br. Heart J. *45:* 411–416 (1981).

Hossack, K.F.; Bruce, R.A.; Trimble, S.; et al.: Improved exercise performance in persons with stable angina pectoris receiving diltiazem. Am. J. Cardiol. *47:* 95–101 (1981).

Hossack, K.F.; Pool, P.E.; Steele, P.; et al.: Efficacy of diltiazem in angina on effort: a multicentre trial. Am. J. Cardiol. *49:* 567–572 (1982).

Hung, J.; Lamb, I.H.; Connolly, S.J.; et al.: The effect of diltiazem and propranolol, alone and in combination, on exercise performance and left ventricular function in patients with stable effort angina: a double-blind randomized and placebo controlled study. Circulation *68:* 560–567 (1983).

Jaiprakash, S.S.; Chatterjee, S.S.; Sahay, J.N.; et al.: Efficacy of nifedipine in the treatment of angina pectoris and chronic airway obstruction. Post-grad. med. J. *56:* 624–628 (1980).

Johnston, D.L.; Lesoway, R.; Humen, D.P.: Clinical and hemodynamic evaluation of propranolol in combination with verapamil, nifedipine and diltiazem in exertional angina pectoris. Am. J. Cardiol. *55:* 680–687 (1985).

Kloner, R.A.; Braunwald, E.: Effects of calcium antagonists in infarcting myocardium. Am. J. Cardiol. *59:* 84B–94B (1987).

Levin, D.C.; Fallon, J.T.: Significance of the angiographic morphology of localized coronary stenoses. Histopathologic correlations. Circulation *66:* 316–320 (1982).

Low, R.I.; Takeda, P.; Lee, G.; et al.: Improved exercise performance in persons with stable angina pectoris receiving diltiazem. Am. J. Cardiol. *47:* 95–101 (1981).

Mandelkorn, J.B.; Wolf, M.N.; Singh, S.; Schechter, J.A.; Kersh, R.I.; Rodgers, D.M.; Workman, M.D.; Bentivoglio, L.G.; LaPorte, S.M.; Meister, S.G.: Intracoronary thrombus in nontransmural myocardial infarction and in unstable angina. Am. J. Cardiol. *52:* 1–6 (1983).

Maseri, A.; Severi, S.; L'Abbate, A.; Pesola, A.: 'Variant angina': one aspect of a continuous spectrum of vasospastic angina. Circulation *55/56:* III–120 (1977).

Maseri, A.: Pathogenetic mechanisms of angina pectoris: expanding views. Br. Heart J. *43:* 648–660 (1980).

Maseri, A.; Chiechia, S.; Kaski, J.C.: Mixed angina pectoris. Am. J. Cardiol. *53:* 30E–33E (1985).

Moise, A.; Theroux, P.; Talymans, Y.; Descoings, B.; Lesperance, J.; Waters, D.D.; Pelletier, G.B.; Bourassa, M.G.: Unstable angina and progression of coronary atherosclerosis. New Engl. J. Med. *309:* 685 (1983).

Moskowitz, R.M.; Piccini, P.A.; Nacarelli, G.V.; et al.: Nifedipine therapy for stable angina pectoris: preliminary results of effects on angina frequency and tredmill exercise response. Am. J. Cardiol. *44:* 811–816 (1979).

Moss, A.J.: Current status of calcium channel blocking drugs after myocardial infarction. Eur. Heart. J. *7:* 31A–34A (1986).

Mueller, H.S.; Chahine, R.A.: Interim report of multicenter double-blind placebo controlled studies of nifedipine in chronic stable angina. Am. J. Med. *71:* 645–657 (1981).

Muller, J.E.; Morrison, J.; Stone, P.H.; Rude, R.E.; Rosner, B.; Roberts, R.; Pearle, D.L.; Turi, Z.G.; Schneider, J.F.; Servas, D.F.; Tate, C.; Scheiner, E.; Sobel, B.E.; Hennekens, C.H.; Braunwald, E.: Nifedipine therapy for patients with threatened and acute myocardial infarction: a randomized, double-blind, placebo-controlled comparison. Circulation *69:* 740–747 (1984a).

Muller, J.E.; Turi, Z.G.; Pearle, D.L.; Schneider, J.F.; Serfas, D.H.; Morrison, J.; Stone, P.H.; Rude, R.E.; Rosner, B.; Sobel, B.E.; Tate, C.; Scheiner, E.; Roberts, R.; Hennekens, C.H.; Braunwald, E.: Nifedipine and conventional therapy for unstable angina pectoris: a randomized, double-blind comparison. Circulation *69:* 728–739 (1984b).

Nayler, W.G.; Horowitz, J.D.: Calcium antagonists: a new class of drugs. Pharmacol. Ther. *20:* 203–262 (1983).

Ouyand, P.; Brinker, J.A.; Mellits, E.D.; Weisfeldt, M.L.; Gerstenblith, G.: Variables predictive of successful medical therapy in patients with unstable angina: selection by multivariate analysis from clinical electrocardiographic, and angiographic evaluations. Circulation *70:* 367–376 (1984).

Parodi, O.; Simonetti, I.; Michelassi, C.; Carpeggiani, C.; Biagini, A.; L'Abbate, A.; Maseri, A.: Comparison of verapamil and propranolol therapy for angina pectoris at rest: a randomized, multiple-crossover, controlled trial in the coronary care unit. Am. J. Cardiol. *57:* 988–996 (1986).

Pine, M.B.; Citron, P.D.; Bailly, D.J.: Verapamil versus placebo in relieving stable angina pectoris. Circulation *65:* 17–22 (1982).

Pool, P.E.; Seagren, S.C.: Long-term efficacy of diltiazem in chronic stable angina associated with atherosclerosis: effect on treadmill exercise. Am. J. Cardiol. *49:* 573–577 (1982).

Pratt, C.M.; Gibson, R.; Boden, W.; Theroux, P.; Strauss, H.; Gheovghiade, M.; Capone, R.; Crawford, M.; Schlant, R.; Arensberg, D.; Kleiger, R.; Perryman, M.B.; Schechtman, K.; Young, Ph.; and the Diltiazem Reinfarction Study Group: Design of a multicenter, double-blind study to assess the effects of prophylactic diltiazem on early reinfarction after non-Q-wave acute myocardial infarction: diltiazem reinfarction study. Am. J. Cardiol. *58:* 906–910 (1986).

Rentrop, K.P.: Thrombolytic therapy in patients with acute myocardial infarction. Circulation *71:* 627–631 (1985).

Rentrop, K.P.; Felt, F.; Blanke, H.; Stecy, P.; Schneider, R.; Roey, M.; Horowitz, S.; Goldman, M.; Karsch, K.; Meilman, H.; Cohen, M.; Siegel, S.; Sanger, J.; Slater, J.; Gorlin, R.; Fox, A.; Fagerstrom, R.; Calhoun, W.F.: Effects of intracoronary streptokinase and intracoronary nitroglycerin infusion and coronary angiographic patterns and mortality in patients with acute myocardial infarction. New Engl. J. Med. *311:* 1457–1463 (1984).

Report of the Holland Interuniversity Nifedipine/Metoprolol Trial (HINT) Research Group: Early treatment of unstable angina in the coronary care unit: a randomised, double-blind, placebo controlled comparison of recurrent ischaemia in patients treated with nifedipine ov metoprolol or both. Br. Heart J. *56:* 400–413 (1986).

Rizzon, P.; Scrutinio, D.; Mangini, S.G.; Lagioia, R.; Toma, L. de: Randomized placebo-controlled comparative study of nifedipine, verapamil and isosorbide dinitrate in the treatment of angina at rest. Eur. Heart J. *7:* 67-76 (1986).

Sherman, L.G.; Chang, S.L.: Nifedipine in chronic stable angina: a double-blind placebo controlled crossover trial. Am. J. Cardiol. *51:* 706–711 (1983).

Sherman, C.T.; Litvack, F.; Grundfest, W.; Lee, M.E.; Chaux, A.; Kass, R.; Swan, H.J.C.; Matloff, J.; Forrester, J.S.: Fiberoptic coronary angioscopy identifies thrombus in all patients with unstable angina (Abstract). Circulation *72:* suppl. II, pp. II–112 (1985).

Simoons, M.L.; Taams, M.; Lubsen, J.; et al.: Treatment of stable angina pectoris with verapamil hydrochloride: a double-blind crossover study. Eur. Heart J. *1:* 269–274 (1980).

Sirnes, P.A.; Overskeid, K.; Pedersen, T.R.; Bathen, J.; Drivenes, A.; Froland, G.S.; Kjekshus, J.K.; Landmark, K.; Rokseth, R.; Sirnes, K.E.; Sundoy, A.; Torjussen, B.R.; Westlund, K.M.; Wik, B.A.: Evolution of infarct size during the early use of nifedipine in patients with acute myocardial infarction: the Norwegian nifedipine multicenter trial. Circulation *70:* 638–644 (1984).

Strauss, W.E.; McIntyre, K.M.; Parisi, A.F.; et al.: Safety and efficacy of diltiazem hydrochloride for the treatment of stable angina pectoris: report of a cooperative clinical trial. Am. J. Cardiol. *49:* 560–566 (1982).

Subramian, B.; Bowles, M.J.; Davies, A.B.; et al.: Combined therapy with verapamil and propranolol in chronic stable angina. Am. J. Cardiol. *49:* 125–132 (1982).

The Danish Study Group on Verapamil in Myocardial Infarction: verapamil in acute myocardial infarction. Eur. Heart J. *5:* 516–528 (1984).

Thuesen, L.; Jorgenson, J.L.; Kvistgaard, H.J.; Sorensen, J.A.; Vaeth, M.; Jensen, E.B.; Jensen, J.J.; Hagerup, L.: Effect of verapamil on enzyme release after early intravenous administration in acute myocardial infarction: double-blind randomized trial. Br. med. J. *286:* 1107–1108 (1983).

Uusitalo, A.; Arstila, M.; Bae, E.A.: Metoprolol, nifedipine and the combination in stable effort angina. Am. J. Cardiol. *57:* 733–737 (1986).

Wagniart, P.; Ferguson, R.J.; Chaitman, B.R.; et al.: Increased exercise tolerance and reduced electrocardiographic ischemia with diltiazem in patients with stable angina pectoris. Circulation *66:* 23–28 (1982).

Wilcox, R.G.; Hampton, J.R.; Banks, D.C.; Birkhead, J.S.; Brooksby, J.S.; Burns-Cox, C.J.; Hayes, M.J.; Joy, M.D.; Malcolm, A.D.; Mather, H.G.; Rowley, J.M.: Trial of earley nifedipine in acute myocardial infarction: The TRENT study. Br. med. J. *293:* 1204–1208 (1986).

Winniford, M.; Fulton, K.L.; Corbett, J.R.: Propranolol-verapamil versus propranolol-nifedipine in severe angina pectoris of effort. Am. J. Cardiol. *55:* 281–285 (1985).

W.H. van Gilst, MD, Department of Cardiology, Oostersingel 59,
NL–9713 EZ Groningen (The Netherlands)

Van Zwieten PA (ed): Clinical Aspects of Calcium Entry Blockers.
Prog Basic Clin Pharmacol. Basel, Karger, 1989, vol 2, pp 103–124

Treatment of Hypertension with Calcium Antagonists

Lennart Hansson, Björn Dahlöf

Department of Medicine, University of Göteborg, Östra Hospital, Göteborg, Sweden

Background

Nomenclature

Calcium ions are essential for the process that couples membrane excitation and the activation of contractile proteins in all muscle cells [1–3]. The concentration of intracellular calcium ions determines the activity of the contractile apparatus [4, 5].

Compounds that interfere with calcium-dependent processes or regulation mechanisms at the cell membrane level and/or intracellularly have usually been called calcium antagonists (CAs), but there have been numerous alternatives such as calcium antagonist (or inhibitor or blocker), calcium entry antagonist (or blocker), slow-channel antagonist (or inhibitor or blocker). In attempts to classify the different compounds within this group, the term 'calcium antagonist' has usually been employed, e.g., by Godfraind [6], and by the World Health Organization Expert Committee on the Classification of Calcium Antagonists [7] and most recently by the International Society and Federation of Cardiology Working Group on the Classification of Calcium Antagonists [8]. Perhaps more important is the fact that Fleckenstein [9], who already in 1964 described that verapamil and prenylamine could mimic the effects of calcium withdrawal on the myocardium, preferentially has used the terminology 'calcium antagonist'. In addition, since it cannot be excluded that some of the therapeutic effects of CAs are exerted through an interaction with the intracellular calcium-transporting protein calmoduline [10] and since the prophylactic actions of CAs against experimental calcinosis [11] appear to be unrelated to the blockade of calcium channels, the more general term 'calcium antagonist' can be justified.

Calcium and Hypertension

Interest for the relationship between calcium and arterial hypertension goes back several decades. In the 1950s and 60s, several studies demonstrated an increased prevalence of hypertension in patients with hyperparathyroidism. A review of 9 studies showed that 10–70% of patients with hyperparathyroidism were hypertensive [12]. Patients with hyperparathyroidism who did not have overt hypertension still had higher blood pressures than matched control subjects [12]. Moreover, numerous studies, e.g., from Japan, USA, Great Britain, Ireland, Canada and Sweden have shown that individuals who live in geographical areas with increased water hardness to some extent are protected against coronary heart disease and cerebrovascular disorders [13]. It can, of course, be argued that increased hardness of water is not only attributable to an increased concentration of calcium, but that increased concentrations of magnesium, zinc, copper, lead, iron, cadmium, lithium and molybdenum may play a role as well [13].

Interest in the calcium ion was renewed when McCarron [14] published a report showing that untreated hypertensive patients had reduced serum concentrations of ionized calcium. In a similar study comprising 98 hypertensive patients there was no overall reduction of ionized serum calcium [15]. However, following a classification of the patients into those with low, normal and high renin activity, it was shown that the concentration of ionized serum calcium was reduced in the low-renin group and increased in the high-renin group as compared to a normotensive control group [15].

The analysis performed by McCarron et al. [16] of the large-scale health and nutrient examination performed in the USA (Health and Nutrition Examination Survey I, HANES I) led to the claim that low dietary calcium was the best nutrient predictor of high blood pressure. However, other investigators, who used the same data but more appropriate corrections for age, etc., could not confirm the important role of calcium in relation to arterial hypertension [17–19]. Still, numerous intervention trials were conducted with the aim of increasing calcium intake and thereby reduce the level of blood pressure. The daily dose of calcium in such studies has varied between 0.4 and 2 g/day in studies comprising between 10 and 90 patients treated for periods varying between 5 days and 4 years [20–28]. In four studies calcium supplementation did not lower blood pressure at all [22, 25, 26, 28]. In a 5-day study of 10 hypertensive patients there was no overall effect on blood pressure; only in the low-renin patients did diastolic blood pressure fall by 6 mm Hg [20]. In an extended study of 34 patients treated for up to 4 years the resting systolic blood pressure fell by 13 mm Hg, but since antihypertensive drugs were given as well it is difficult to assess the contribution of the calcium supplementation [21]. In studies by McCarron et al. [23] and Morris et al. [27] significant effects have been reported regarding standing systolic blood pressure, in the order of 5.6–4 mm Hg, whereas supine and diastolic blood pressures were affected less or not at all [23, 27]. In a study of 90 hypertensive patients treated for 12 weeks there was a temporary reduction in sitting diastolic blood pressure by 3.1 mm Hg after 6 weeks, but after 12 weeks of calcium supplementation there was no demonstrable effect [24]. Based on these clinical observation, it seems fair to conclude that it is not justified to recommend increased calcium intake as a means of lowering elevated arterial pressure. The relationship between calcium, blood pressure and renin is interesting, however, and based on our present knowledge the following summary can be made: patients with low-renin hypertension have reduced ionized serum calcium, patients with high-renin hypertension have increased ionized serum calcium and CAs appear to exert a more potent antihypertensive effect in patients with low-renin hypertension [15].

Early Use of Calcium Antagonists in Hypertension

Already in 1970, two reports were published in which intravenous verapamil was shown to have an antihypertensive effect when used in hypertensive emergencies [29, 30]. Later, oral administration of CAs were shown to lower elevated blood pressure, e.g., by pioneers such as Lederballe Pedersen [31] and Aoki et al. [32]. During the 1980s, progress has been extremely fast as regards the development of new CAs for the treatment of cardiovascular disorders, including arterial hypertension. It would be impossible to give a balanced review of all the data which have contributed to the present status of CAs, when many authorities hold the view that this class of compounds should be considered as one of the alternatives for first-line treatment in an individualized therapeutic approach to the treatment of hypertension [33, 34].

Chemistry – Pharmacology – Pharmacokinetics

The chemistry, pharmacology and pharmacokinetics of CAs have been covered in chapter 1 of this book. Suffice to say that a recently proposed classification of CAs includes three groups, in which group A consists of highly specific CAs, mainly verapamil, dihydropyridine derivatives and diltiazem, which all have separate, although slightly overlapping, binding sites (table I) [8].

Group B includes less specific agents, e.g., agents with combined Na^+/Ca^{++} channel blockade, which may or may not inhibit the nitrendipine ligand binding (table I). Finally, group C includes nonspecific agents, which may possess significant calcium antagonism only as a side effect, e.g., following excessive overdosing.

Of particular interest for the treatment of hypertension is the fact that different CAs show great variations in their relative affinity for myocardial cells as compared to vascular smooth muscle cells. Thus, in one study, the in vitro affinity of verapamil for myocardial vs. vascular muscle cells was approximately 1:1, whereas nifedipine had a relative selectivity in the order of 1:10 and felodipine of 1:100 [35]. Since virtually all forms of established hypertension are characterized by increased total peripheral vascular resistance and normal or low cardiac output [36], it is obvious that a CA with a vascular profile would appear preferable for the treatment of hypertension, whereas agents with predominantly cardiac effects would be chosen, e.g., in the treatment of supraventricular arrhythmias.

Effects in Hypertension

Hemodynamics

All CAs are coronary vasodilators and all compounds developed so far have a vasodilating action if they are given in sufficiently high doses [1]. Obviously, the degree of vasodilatation in relation to the myocardial effects will differ, depending on the degree of cardiac/vascular selectivity, briefly discussed above. In general, the vascular smooth muscle cells of the arterioles are far more sensitive to treatment with CAs than those in the veins [37]. It appears that the dihydropyridine derivative, nisoldipine, may be an exception, since it has been reported that this agent has a powerful venodilator effect [38]. Moreover, vascular effects are greater in hypertensive than in normotensive animals [39].

The acute hemodynamic effects of CAs consist of a rapid drop in blood pressure, usually due to a fall in vascular resistance, particularly when dihydropyridine derivatives have been used. Thus, in one study Guazzi et al. [40] found a maximal reduction in blood pressure within 30 min of

Table I. Classification of CAs[1]

Group A

1 Major mode of action is blockade of voltage-dependent calcium channels
2 Verapamil, nifedipine (and related dihydropyridines) and diltiazem are the prototypes
3 Three binding sites: verapamil, nifedipine and diltiazem: there is a certain overlap, especially between verapamil and diltiazem

Group B

1 Less specific agents with inhibition of the nitrendipine ligand binding:
a Combined Na^+/Ca^{2+} channel blockers (bepridil, tiapamil, fendiline)
b Cinnarizine and flunarizine (binding sites separate from verapamil, nifedipine and diltiazem)
2 Less specific agents without inhibition of the nitrendipine ligand binding, e.g., prenylamine, perhexiline and others

Group C

Nonspecific agents with calcium antagonism as a side effect, e.g., following substantial overdosage, chlorpromazine and others

[1] Proposed by the International Federation and Society of Cardiology Working Group on Classification of Calcium Antagonists for Cardiovascular Diseases [8].

giving a single oral dose of nifedipine and 20 min after giving the same dose sublingually. The fall in mean arterial blood pressure was 21%, which could be attributed to a marked reduction in total peripheral vascular resistance (39%), whereas cardiac output and heart rate increased [40].

During long-term treatment with long-acting nifedipine a sustained fall in intra-arterial blood pressure was demonstrated which, even after 1 year of treatment, was entirely attributable to a fall in total peripheral resistance whereas cardiac index and heart rate were unchanged [41] (fig. 1). The hemodynamic results of verapamil treatment are less consistent. The fall in arterial pressure has been reported by some authors to coincide with an increase in heart rate [42, 43], whereas others have found no change [44–46] or even a fall in resting heart rate [47–49]. Similarly, cardiac output has been shown to increase [44, 50], to remain unchanged [46, 51] or to decrease [52, 53].

These discrepancies regarding the hemodynamic effects of verapamil may be attributable to the fact that different populations have been investigated and that different doses have been used. However, the fact that verapamil has a complex mode of action, which results in negative

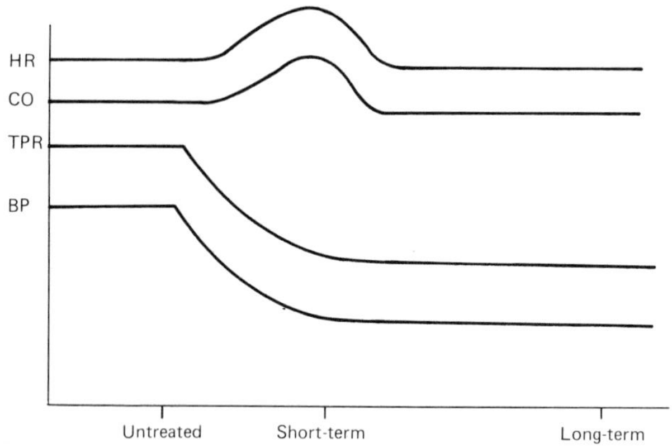

Fig. 1. Schematic illustration of the hemodynamic changes induced during short-term and long-term antihypertensive treatment with CAs of the dihydropyridine type. HR = Heart rate; CO = cardiac output; TPR = total peripheral resistance; BP = blood pressure.

chronotropism and inotropism, coronary and peripheral vasodilatation and reflex sympathetic stimulation may contribute to these inconsistencies [54].

Diltiazem reduces blood pressure acutely after intravenous administration and this coincides with a temporary rise in heart rate which is not sustained, whereas systemic vascular resistance falls significantly [55].

Nifedipine, verapamil and diltiazem have all been found to affect pulmonary vasoconstriction during experimental pulmonary hypertension [56]. However, no effect of diltiazem was shown acutely on hypoxic pulmonary vasoconstriction in patients with chronic hypoxia [57].

Peripheral hemodynamic studies have shown that both verapamil and nifedipine reduce forearm vascular resistance acutely [58, 59]. Similar observations on forearm and digital blood flow have been made after acute administration of diltiazem [60].

Mode of Action

As briefly discussed above, most CAs appear to lower blood pressure mainly through vasodilatation, although verapamil and possibly also diltiazem to a certain extent may in addition reduce cardiac output during chronic administration. The mechanisms underlying the vasodilator activity of CAs appear to be mainly through interference with receptor-operated calcium channels [1]. Recent investigations indicate that both α_1- and α_2-adrenoceptors may play a role in this context [61]. Moreover, interesting new data suggest that interference with angiotensin II in the vascular wall may play a role as well and that the transmembranous calcium influx may depend on receptor-operated calcium channels, triggered by either α-adrenoceptors or angiotensin II receptors [61].

In addition, it cannot be ruled out that the weak natriuretic and diuretic effect possessed by several CAs may contribute to the antihypertensive effect, mainly by preventing fluid and salt retention, which frequently causes secondary drug resistance with pure vasodilators such as hydralazine and minoxidil [62].

The first clinical observation of an acute increase in sodium excretion and urinary volume after administration of nifedipine was made in 1972 [63]. Similar results have been obtained in normotensive subjects both with nifedipine and nitrendipine [64]. Studies carried out in experimental animals and in hypertensive patients and normotensive subjects with, e.g., verapamil [65, 66], diltiazem [67] and felodipine [68, 69] by and large confirm the acute natriuretic and diuretic effects.

Although there are few direct comparisons between different agents, it appears that the dihydropyridine derivatives may have more potent diuretic and natriuretic effects than, e.g., verapamil and that these effects are more pronounced in hypertensive patients than in normotensive subjects. The exact mechanisms behind these effects are not known, but renal micropuncture studies show that the reabsorption of sodium is reduced in the distal renal tubules [70].

Antihypertensive Effect

Long-Term. A multitude of studies have shown that CAs have a useful antihypertensive effect. Thus, nifedipine has been shown to be equally as potent as bendroflumethiazide [71], propranolol [32], metoprolol [72], atenolol [73] and labetalol [74]. Similarly, verapamil has been found to be as effective in lowering elevated arterial pressure as propranolol [75–77], pindolol [78], α-methyldopa [75], reserpine [79] and nifedipine [80, 81].

Diltiazem was found to be as effective as propranolol in essential hypertension and it lowered blood pressure more effectively during submaximal exercise [82]. Diltiazem and nifedipine have been shown to be equipotent when compared in 40 hypertensive patients [83].

In combined treatment with other agents an increased antihypertensive effect has been demonstrated, e.g., for nifedipine + a diuretic [71], nifedipine + labetalol [74] and the combination of prazosin and verapamil has been reported to be extremely potent, presumably due to an interaction between these two agents [84].

Still, the clinically most interesting combination appears to be between a β-blocking agent and a dihydropyridine-type CA. Thus, in a comparison with hydralazine, felodipine given as added treatment to β-blockers in 120 patients was markedly more effective than hydralazine while being better tolerated [85]. Other studies confirm the great efficacy of felodipine when used together with β-blockers [86, 87].

Isradipine used in combination with pindolol has been shown to be markedly potent, bringing the blood pressure down to strictly normotensive levels in a majority of patients [88] (fig. 2). Also combinations between metoprolol and nifedipine or nitrendipine have shown great efficacy [89].

Another combination which appears to have great potential is that of a CA and an angiotensin-converting enzyme inhibitor [90]. Based mainly on favorable clinical impressions, several prospective studies are now investigating this interesting although costly therapeutic combination.

It has been shown that the antihypertensive effect of CAs correlates significantly with the age of the patient, so that elderly hypertensives respond better [91]. Other investigators have found much weaker or nonsignificant correlations between age and the antihypertensive response [88, 41]. From a clinical point of view it is clear that treatment of hypertension with CAs does not need to be reserved for the elderly.

Plasma renin activity has been used as a predictor of the antihypertensive response. Several studies have shown that patients with low plasma renin activity respond better to CAs, e.g. [91]. The predictive value is relatively low, however, and from a clinical point of view it does not appear to be justified to do renin studies on this ground.

Finally, numerous studies have shown that the antihypertensive response to CAs is better the higher the initial untreated blood pressure and several investigators have presented statistical correlations to prove this point. No attempt will be made here to review this topic since such correlations are meaningless from a clinical point of view and not correct from a statistical point of view.

Short-Term. Several studies have shown that nifedipine given sublingually, e.g., 5 mg, will cause a rapid reduction in arterial pressure. The onset of action is within 5 min and this approach can therefore be used in

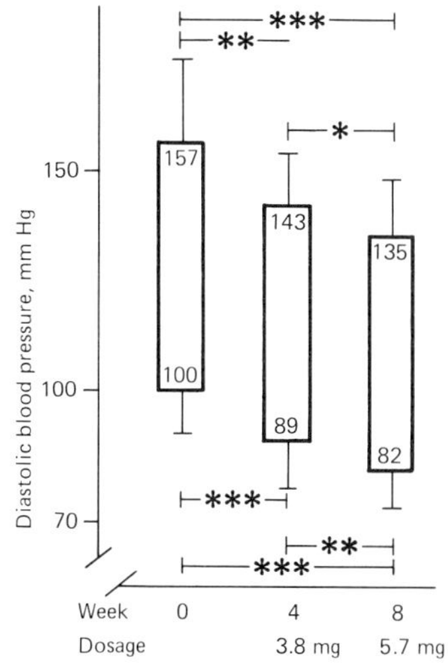

Fig. 2. Marked augmentation of the antihypertensive effect when the dihydropyridine CA isradipine (PN 200–110) was added to previous treatment with pindolol in 20 patients with essential hypertension. With the addition of a dosage of isradipine (5.7 mg b.i.d.), average diastolic blood pressure was reduced by 18 mm Hg [88].

hypertensive emergencies, with the added benefit that cerebral blood flow is not reduced, which is in contrast, e.g., to treatment with clonidine [92]. Still, it must be kept in mind that acute and drastic lowering of blood pressure in hypertensive emergencies should only be attempted on strict indications [62].

Effects on Cardiovascular Structure

It has been suggested that a more ambitious aim with antihypertensive treatment would be not only to reduce elevated blood pressure levels but also to obtain regression of hypertension-induced cardiovascular hypertrophy [93]. Regression of left-ventricular hypertrophy has been shown with nifedipine [94], felodipine [95] and diltiazem [96]. It is conceivable that this is a class-effect that should be expected with all calcium antagonists since the long-term antihypertensive effect is obtained without significant reflex sympathetic activation or other indications of increments of trophical factors.

Regression of structural arteriolar changes has been obtained in animal studies with felodipine [95, 97] although the results have not been uniform in this respect [98]. In combination with metoprolol, it has been shown to induce a reversal of structural coronary hypertrophy [99].

Additional Effects

Atherosclerotic complications are common in hypertension. Therefore, some observations that treatment with nifedipine may suppress atherogenesis in cholesterol-fed rabbits [100] are of great potential clinical interest. So are the observations that treatment with CAs may prevent vascular calcium overload, a known pathogenic factor in arteriosclerosis [101]. Particularly the work by Fleckenstein et al. [101] and Fleckenstein [102] is promising, since they have shown that arterial calcinosis can be prevented by treatment with CAs. Animal studies with isradipine have shown that this interesting effect can be obtained with doses that are in the clinical range [103, 104].

Adverse Effects

The adverse effects caused by CAs differ depending on the class of compound used. The dihydropyridine derivatives, which have a strong vasodilator effect, are more likely to cause flushing, sensation of heat, palpitations, tachycardia, dizziness and vertigo [105–108]. Deterioration of cardiac function has more frequently been reported with verapamil [105, 106]. Particularly, the combination between a CA of the verapamil type and a β-blocker may cause problems in this respect. The cardiovascular adverse effects of combined CA and β-blocker therapy have been reviewed separately [109]. In brief it can be said that verapamil used with a β-blocker was associated with conduction problems in up to 9% and dyspnea or heart failure in up to 8%, whereas in combinations with nifedipine ankle edema, flushing and headaches predominated [109].

The dihydropyridine derivatives frequently cause edema around the ankles [59, 85, 86, 109]. Obviously, this is due to local precapillary vasodilatation, which in combination with upright posture increases the transcapillary filtration pressure and, as implied in the discussion above about the natriuretic and diuretic effects of these agents, the edema is not an expression of general fluid retention.

Gastrointestinal symptoms such as nausea and constipation are quite frequent with verapamil [108], whereas nifedipine may cause diarrhea [110].

Table II. Side effects caused by CAs[1]

	Verapamil	Nifedipine	Diltiazem
Cardiovascular system			
Hypotension	+	+	+
Flush	+	++	–
Headaches	+	++	+
Ankle oedema	+	+	+
Palpitations/chest pain	+	+	+
A-V conduction disturbance	++	–	+
Heart failure	+	(+)	–
Bradycardia	++	–	+
Gastrointestinal system			
Nausea	+	–	+
Constipation	+	–	(+)
Diarrhea	–	+	+

[1] Modified after Krebs [107].

Some of the important side effects caused by the three major classes of CAs are outlined in table II [110].

Metabolic side effects are rarely of clinical significance during treatment with CAs. Trost and Weidmann [111] made an extensive review of the literature concerning CAs and glucose metabolism in man and concluded that CAs do not have clinically relevant effects on glucose homeostasis in diabetic or non-diabetic individuals. It has been shown that long-term treatment with nitrendipine in diabetic patients did not affect Hb_{A1} levels, weight or the antidiabetic therapy during 6 months of treatment [112].

Moreover, neither nifedipine nor verapamil has any noticeable effect on serum lipoproteins [113].

Interactions

Interactions between CAs and other pharmacological agents have been reviewed in detail on pp. 1–20. Of immediate interest for the treatment of cardiovascular disorders is the interaction between a β-blocker and CAs of the verapamil type. As noted above, this combination may cause A-V conduction problems in up to 9% of patients and dyspnea or heart failure [109]. Profound cardiac failure, hypotension and bradycardia may occur in patients with ischemic heart disease, even if their baseline left-ventricular function is normal or only mildly reduced [114]. For such reasons, it must be regarded as sound clinical practice to avoid this combination, even if it is not absolutely contraindicated. On the other hand, it has been stated that dihydropyridine-derived CAs may be used safely in combination with β-blockers [105, 106].

The relevant pharmacokinetic interaction between prazosin and verapamil has already been discussed briefly above [84]. Since this combination of drugs may cause severe hypotension, great care should be observed.

Finally, serum digoxin levels may increase after administration of verapamil [115–118] or nifedipine [115]. The effect of verapamil in this respect is dose-dependent and greater than that of nifedipine [115]. However, it is doubtful that this effect is clinically relevant or whether it necessitates an adjustment of the digoxin dosage [117].

Indications

As discussed above, the antihypertensive effect of CAs when used in the treatment of hypertension has been amply documented. There is no doubt that the dihydropyridine-derived CAs almost without exception can be used in the place of pure vasodilators such as hydralazine or minoxidil. However, many authorities, e.g., Zanchetti [33], have proposed a more individualized approach to the drug treatment of hypertension, which would open up the possibility of using CAs as first-line treatment (fig. 3). Moreover, Erne et al. [34] have advocated a more frequent use of CAs, particularly in the treatment of hypertension in the elderly.

Since CAs of all the three major classes may be used successfully in the treatment of angina pectoris, it is obvious that these drugs are well suited for the management of patients with coexisting angina pectoris and hypertension [119]. If the observations of the antiatherosclerotic and anticalcinotic effects, briefly discussed above, are confirmed in clinical trials, it is obvious that these agents will find indications in the treatment of patients with hypertension and arteriosclerotic complications.

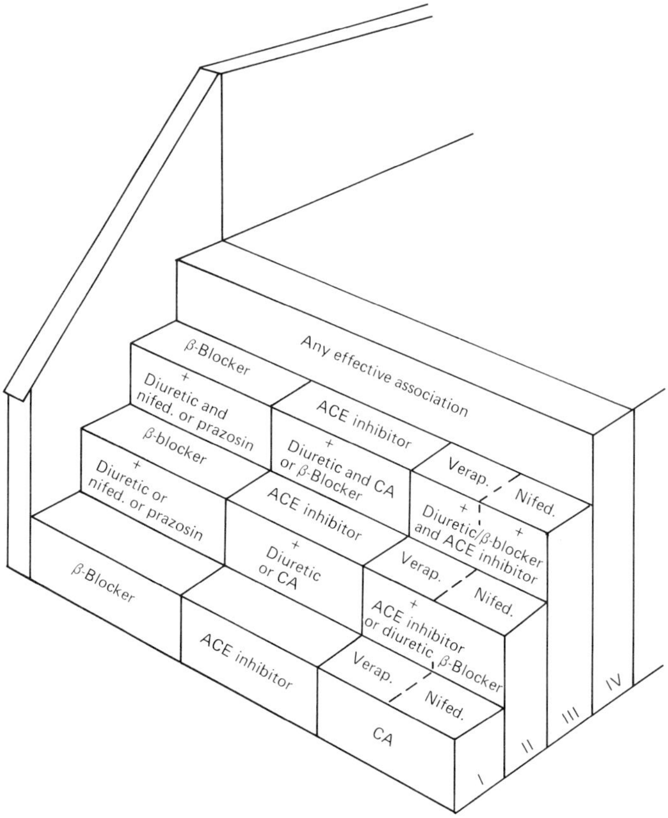

Fig. 3. Proposal for a revised stepped-care approach to the treatment of hypertension, in which CAs could be used as first-line therapy. From Zanchetti [33], with kind permission from the author and the editor.

Contraindications

There are few absolute contraindications to the use of CAs in hypertension. However, verapamil and diltiazem should not be given to patients with sick sinus syndrome or with A-V conduction disturbances of grades II–III. Great care should also be exercised when giving these agents together with a β-blocker, even to patients with apparently normal left-ventricular function [109, 114].

The dihydropyridine derivatives should not be given to patients with untreated left-ventricular failure or cardiogenic shock.

Socio-Economic Aspects

Compliance

Compliance to antihypertensive treatment is affected by many different factors, in particular the number and severity of adverse effects in relation to the expected or experienced therapeutic effect [93]. Another important factor is that treatment should not negatively affect the patient's quality of life [120].

So far, the quality of life during treatment with CAs has not been evaluated as extensively as, e.g., the effect of angiotensin-converting enzyme inhibitors on well-being [121]. However, with the exception of patients who suffer subjective adverse effects, there is little reason to assume that any of the CAs will have negative effects on well-being. Needless to say, this important topic warrants an objective and conclusive evaluation.

Cost

It is difficult to make a realistic estimate of the cost of antihypertensive treatment. A simplified way of looking at this is to compare the cost of pills and in such a comparison, CAs obviously appear more expensive than, e.g., thiazide diuretics. However, there is no need to monitor clinical chemistry variables such as serum potassium and serum urate during treatment with CAs and this shifts the balance slightly when CAs and thiazides are compared. The more important issue is, of course, to evaluate the cost of treatment in relation to the expected or demonstrated benefits on morbidity and mortality. Since no intervention trials have been conducted with CAs with the aim to study morbidity and mortality, this question cannot be answered at present. It must be assumed, though, that the benefits obtained through treatment with CAs, as with all other antihypertensive agents, far outweigh the costs.

Future Aspects

It appears likely that the present strong position of CAs in the treatment of hypertension will be further reinforced in the future. Several factors may contribute to this, e.g., the development of more potent and longer-acting agents, probably mainly within the dihydropyridine derivative group. Other factors that may contribute to the strengthening of CAs in the antihypertensive armamentarium would be a clear demonstration of clinical benefits derived from the antiatherosclerotic and decalcifying effects briefly discussed above. Finally, if a more ambitious therapeutic aim in hypertension is to be sought, e.g., normotensive blood pressure levels in most patients, it is obvious that antihypertensive agents will have to be combined more frequently than has been common so far [122]. In this perspective, combinations between dihydropyridine-derived CAs and β-blockers offer an attractive alternative and possibly the combination between a CA and an angiotensin-converting enzyme inhibitor will provide another equally effective alternative. Observations that pretreatment with an angiotensin-converting enzyme inhibitor may alter the response of certain vascular beds from vasoconstriction or no response to vasodilatation [123] provide theoretical support of this view.

References

1 van Zwieten, P.A.: Calcium entry blockers (calcium antagonists); in van Zwieten, Handbook of hypertension; vol. 3, Pharmacology of antihypertensive drugs, pp. 331–342 (Elsevier, Amsterdam 1984).

2 Andersson, K.-E.: Effects of calcium and calcium antagonists on the excitation-contraction coupling in striated and smooth muscle. Acta pharmac. Tox. *43:* suppl. 1, pp. 5–14 (1978).

3 Stull, J.T.; Sanford, C.F.: Differences in skeletal, cardiac and smooth muscle contractile element regulation by calcium; in Weiss, New perspectives on calcium antagonists, pp. 35–43 (American Physiological Society, Bethesda 1981).

4 Ebashi, S.: Regulation of muscle contraction. Proc. R. Soc. Lond. *207:* 259 (1980).

5 Blaustein, M.P.: Sodium ions, calcium ions, blood pressure regulation, and hypertension: a reassessment and a hypothesis. Am. J. Physiol. *232:* 165–173 (1977).

6 Godfraind, T.: Classification of calcium antagonists. Am. J. Cardiol. *59:* 11B–23B (1987).

7 Vanhoutte, P.M.: The Expert Committee of the World Health Organization on Classification of Calcium Antagonists: the viewpoint of the rapporteur. Am. J. Cardiol. *59:* 3A–8A (1987).

8 Opie, L.H.; Bühler, F.R.; Fleckenstein, A.; Hansson, L.; Harrison, D.C.; Poole-Wilson, P.A.; Schwartz, A.; Vanhoutte, P.M.: International Society and Federation of Cardiology classification of calcium antagonists for cardiovascular diseases. Am. J. Cardiol. (in press).

9 Fleckenstein, A.: Die Bedeutung der energiereichen Phosphate für Kontraktilität und Tonus des Myokards. Verh. dt. Ges. inn. Med. *70:* 81–99 (1964).

10 Broström, S.L.; Ljung, B.; Mårdh, S.; Forsén, S.; Thulin, E.: Interaction of the antihypertensive drug felodipine with calmoduline. Nature *292:* 777–778 (1981).

11 Frey, M.; Keidel, J.; Fleckenstein, A.: Verhütung experimenteller Gefäß-Verkalkungen (Mönckeberg's Typ der Arteriosklerose) durch Calcium-Antagonisten; in Fleckenstein, Roskamm, Calcium Antagonismus, pp. 258–264 (Springer, Berlin 1980).
12 Christensson, T.; Hellström, K.; Wengle, B.: Blood pressure in subjects with hypercalcaemia and primary hyperparathyroidism detected in a health screening programme. Eur. J. clin. Invest. *7:* 109–113 (1977).
13 Kesteloot, H.: Blood pressure, calcium and water hardness; in Bulpitt, Handbook of hypertension, vol. 6, Epidemiology of hypertension, pp. 216–229 (Elsevier, Amsterdam 1985).
14 McCarron, D.A.: Low serum concentrations of ionized calcium in patients with hypertension. New. Engl. J. Med. *307:* 226–228 (1982).
15 Resnick, L.M.; Laragh, J.H.; Sealey, J.E.; Alderman, M.H.: Divalent cations in essential hypertension. Relations between serum ionized calcium, magnesium and plasma renin activity. New. Engl. J. Med. *309:* 888–891 (1983).
16 McCarron, D.A.; Morris, C.D.; Henry, H.J.; Stanton, J.L.: Blood pressure and nutrient intake in the United States. Science *224:* 1392–1398 (1984).
17 Feinleib, N.; Lenfant, C.; Miller, S.A.: Hypertension and calcium. Science *226:* 384–386 (1984).
18 Harlan, W.R.; Hull, A.L.; Schmouder, R.L.; Landis, J.R.; Thompson, F.F.; Larkin, F.A.: Blood pressure and nutrition in adults: the National Health and Nutrition Council Examination Survey. Am. J. Epidemiol. *120:* 17–27 (1984).
19 Gruchow, H.W.; Sobocinski, K.A.; Barboriak, J.J.: Alcohol, nutrient intake, and hypertension in US adults. J. Am. Med. Ass. *253:* 1567–1570 (1985).
20 Resnick, L.M.; Laragh, J.H.: The hypotensive effect of short-term oral calcium loading in essential hypertension. Clin. Res. *31:* 334A (1983).
21 Johnson, N.E.; Smith, E.L.; Freudenheim, J.L.: Effects on blood pressure of calcium supplementation of women. Am. J. clin. Nutr. *42:* 12–17 (1985).
22 Singer, D.R.J.; Markandu, N.D.; Cappuccio, F.P.; Beynon, G.W.; Shore, A.C.; Smith, S.J.; MacGregor, G.A.: Does oral calcium lower blood pressure: a double-blind study. J. Hypertens. *3:* 661 (1985).
23 McCarron, D.A.; Morris, C.D.: Blood pressure response to oral calcium in persons with mild to moderate hypertension. A randomized, double-blind, placebo-controlled crossover trial. Ann. intern. Med. *103:* 825–831 (1985).
24 Grobbee, D.E.; Hoffman, A.: Effect of calcium supplementation on diastolic blood pressure in young people with mild hypertension. Lancet *ii:* 703–706 (1986).
25 Nowson, C.; Morgan, T.: Effect of calcium carbonate on blood pressure. J. Hypertens. *4:* suppl. 6, pp. S673–S675 (1986).
26 Zoccali, C.; Mallamaci, F.; Delfino, D.; Ciccarelli, M.; Parlongo, S.; Iellamo, D.; Moscato, D.; Maggiore, Q.: Long-term oral calcium supplementation in essential hypertension: a double-blind, randomized, corss-over study. J. Hypertens. *4:* suppl. 6, pp. S676–S678 (1986).
27 Morris, C.; McCarron, D.A.: Calcium supplementation reduces blood pressure in older systolic hypertensives (Abstract 510). Proc. 11th Scient. Meet. Int. Soc. Hypertens., Heidelberg 1986.
28 Cappuccio, F.P.; Markandu, N.D.; Singer, D.R.J.; Smith, S.J.; Shore, A.C.; MacGregor, G.A.: Does oral calcium supplementation lower high blood pressure? A double-blind study. J. Hypertens. *5:* 67–71 (1987).

29 Brittinger, W.D.; Schwarzbeck, A.; Wittenmeier, K.W.; Twittenhof, W.D.; Stegaru, B.; Huber, W.; Ewald, R.W.; von Henning, G.E.; Fabricius, M.; Strauch, M.: Klinisch-experimentelle Untersuchungen über die blutdrucksenkende Wirkung von Verapamil. Dt. med. Wschr. *95:* 1871–1877 (1970).
30 Bender, F.: Die Behandlung der tachykarden Arrhythmien und der arteriellen Hypertonie mit Verapamil. Arzneimittel-Forsch. *20:* 1310–1316 (1970).
31 Lederballe-Pedersen, O.: Calcium blockade as a therapeutic principle in arterial hypertension. Acta pharmac. tox. *49:* suppl. II, pp. 1–31 (1981).
32 Aoki, K.; Kondo, S.; Mochizuki, A.; Yoshida, T.; Kato, S.; Kato, K.; Takikawa, K.: Antihypertensive effect of cardiovascular Ca^{2+}-antagonist in hypertensive patients in the absence and presence of beta-adrenergic blockade. Am. Heart J. *96:* 218–226 (1978).
33 Zanchetti, A.: A re-examination of stepped-care. A retrospective and prospective. J. cardiovasc. Pharmacol. *7:* suppl. 1, S126–S131 (1985).
34 Erne, P.; Bolli, P.; Bertel, O.; Hulthén, U.L.; Kiowski, W.; Muller, F.B.; Bühler, F.R.: Factors influencing the hypotensive effects of calcium antagonists. Hypertension *5:* suppl. II, pp. II97–II102 (1983).
35 Ljung, B.: Vascular selectivity of felodipine. Drugs *29:* suppl. 2, pp. 46–58 (1985).
36 Freis, E.D.: Hemodynamics in hypertension. Physiol. Rev. *40:* 27–54 (1960).
37 Braunwald, E.: Mechanism of action of calcium-channel-blocking agents. New. Engl. J. Med. *307:* 1618–1627 (1982).
38 Opie, L.H.: Calcium antagonists. Mechanisms, therapeutic indications and reservations: a review. Q. Jl. Med., new Ser. *53:* 1–16 (1984).
39 Masuyama, Y.; Tsuda, K.; Kuchii, M.; Nishio, I.: Effects of diltiazem (a calcium antagonist) on neurosecretion and vascular responsiveness in hypertension. J. Hypertens. *3:* suppl. 3, pp. S169–S172 (1985).
40 Guazzi, M.; Fiorentini, S.; Olivari, M.; Bartorelli, C.; Necci, C.; Polese, A.: Short- and long-term efficacy of a calcium-antagonistic agent (nifedipine) combined with methyldopa in the treatment of severe hypertension. Circulation *61:* 913–919 (1980).
41 Lund-Johansen, P.; Omvik, P.: Haemodynamic effects of nifedipine in essential hypertension at rest and during exercise. J. Hypertens. *1:* 159–163 (1983).
42 Atterhög, J.-H.; Ekelund, L.G.: Haemodynamic effects of intravenous verapamil at rest and during exercise in subjectively healthy middle-aged men. Eur. J. clin. Pharmacol. *8:* 317–322 (1975).
43 Vincenzi, M.; Allegri, P.; Gabaldo, S.; Maiolino, P.; Ometto, R.: Haemodynamic effects caused by intravenous administration of verapamil in healthy subjects. Arzneimittel-Forsch. *26:* 1221–1223 (1976).
44 Ferlintz, J.: Effects of verapamil on normal and abnormal ventricular functions in patients with ischaemic heart disease; in Zanchetti, Krikler, Calcium antagonism in cardiovascular therapy: experience with verapamil, pp. 92–105 (Excerpta Medica, Amsterdam 1981).
45 Sandler, G.; Clayton, G.A.; Thornicroft, S.G.: Clinical evaluation of verapamil in angina pectoris. Br. med. J. *iii:* 224–227 (1968).
46 Singh, B.; Roche, A.: Effects of intravenous verapamil on hemodynamics in patients with heart disease. Am. Heart J. *94:* 593–599 (1977).
47 Andreasen, F.; Boye, E.; Christoffersen, E.; Dalsgaard, P.; Henneberg, E.; Kallenbach, A.; Ladefogel, S.; Lillquist, K.; Mikkelsen, E.; Nordero, E.; Olsen, J.; Pedersen,

J.K.; Pedersen, V.; Petersen, G.; Schroll, J.; Schultz, H.; Seidelin, J.: Assessment of verapamil in the treatment of angina pectoris. Eur. J. Cardiol. *2:* 443–452 (1975).

48 Frishman, W.H.; Klein, N.A.; Strom, J.A.; Willens, H.; LeJemtel, T.H.; Jentzer, J.; Siegel, L.; Klein, P.; Kirschen, N.; Silverman, R.; Pollack, S.; Doyle, R.; Kirsten, E.; Sonnenblick, E.H.: Superiority of verapamil to propranolol in stable angina. A double-blind randomized placebo-controlled cross-over comparison. Circulation *65:* suppl. I, pp. I-152–I-159 (1982).

49 Livesly, B.; Catley, P.F.; Campbell, R.C.; Oram, S.: Double-blind evaluation of verapamil, propranolol, and isosorbide dinitrate against placebo in the treatment of angina pectoris. Br. med. J. *i:* 375–378 (1973).

50 Rydén, L.; Saetre, H.: The haemodynamic effect of verapamil. Eur. J. clin. Pharmacol. *3:* 153–157 (1971).

51 Seabra-Gomes, R.; Rickards, A.; Sutton, R.: Haemodynamic effects of verapamil and practolol in man. Eur. J. Cardiol. *4:* 79–85 (1976).

52 Mangiardi, L.M.; Hariman, R.J.; McAllister, R.G.Jr.; Bhargava, V.; Suravicz, B.; Shabetai, R.: Electrophysiologic and hemodynamic effects of verapamil. Correlation with plasma drug concentrations. Circulation *57:* 366–372 (1978).

53 Ross, G.; Jorgenson, C.R.: Cardiovascular action of iproveratril. J. pharmac. exp. Ther. *158:* 504–509 (1967).

54 Spivack, C.; Ocken, S.; Frishman, W.H.: Calcium antagonists. Clinical use in the treatment of systemic hypertension. Drugs *25:* 154–177 (1983).

55 Joyal, M.; Pieper, J.; Cremer, K.; Feldman, R.L.; Pepine, C.J.: Pharmacodynamic aspects of intravenous diltiazem administration. Am. Heart J. *111:* 54–61 (1986).

56 Young, T.E.; Lundquist, L.J.; Chesler, E.; Weir, E.K.: Comparative effects of nifedipine, verapamil, and diltiazem on experimental pulmonary hypertension. Am. J. Cardiol. *51:* 195–200 (1983).

57 Clozel, J.P.; Delorme, N.; Battistella, P.; Breda, J.L.; Polu, J.M.: Hemodynamic effects of intravenous diltiazem in hypoxic pulmonary hypertension. Chest *91:* 171–175 (1987).

58 Lederballe Pedersen, O.: Does verapamil have a clinically significant antihypertensive effect? Eur. J. clin. Pharmacol. *13:* 21–24 (1978).

59 LederballePedersen, O.; Mikkelsen, E.: Acute and chronic effects of nifedipine in arterial hypertension. Eur. J. clin. Pharmacol. *14:* 375–381 (1978).

60 Finch, M.B.; Johnston, G.D.: The peripheral vascular effects of diltiazem – dose-response characteristics. Br. J. clin. Pharmacol. *20:* 447–451 (1985).

61 van Zwieten, P.A.; Timmermans, P.B.M.W.M.; van Heiningen, P.N.M.; Receptor subtypes involved in the action of calcium antagonists (in press).

62 Hansson, L.: Drug treatment of hypertension; in Robertson, Handbook of hypertension, vol. 1, Clinical aspects of essential hypertension, pp. 397–436 (Elsevier, Amsterdam 1983).

63 Klütsch, K.; Schmidt, P.; Grosswendt, J.: Der Einfluß von Bay a 1040 auf die Nierenfunktion des Hypertonikers. Arzneimittel-Forsch. *22:* 377–380 (1972).

64 Ene, M.D.; Williamson, P.J.; Roberts, C.J.C.; Wadell, G.: Natriuretic effect of nifedipine. Br. J. clin. Pharmacol. *17:* 193 (1984).

65 Dietz, J.R.; Davis, J.O.; Freeman, R.H.; Villareal, D.; Echtenkamp, S.F.: Effects of intrarenal infusion of calcium entry blockers in anesthetized dogs. Hypertension *5:* 482–488 (1983).

66 Leonetti, G.; Sala, C.; Bianchini, C.; Terzoli, L.; Zanchetti, A.: Antihypertensive and renal effects of orally administered verapamil. Eur. J. clin. Pharmacol. *18:* 375–382 (1980).

67 Yokoyama, S.; Mori, N.; Shingu, T.; Sakata, K.; Iwase, T.; Yoshida, H.; Takayama, S.; Hoshino, T.; Kaburagi, T.: Clinical effects of intravenous diltiazem hydrochloride on renal hemodynamics. J. cardiovasc. Pharmacol. *9:* 311–316 (1987).

68 Nordlander, M.; DiBona, G.; Ljung, B.; Yao, T.; Thorén, P.: Renal and cardiovascular effects of acute and chronic administration of felodipine to SHR. Eur. J. Pharmacol. *113:* 25–36 (1985).

69 Edgar, B.; Bengtsson, B.; Elmfeldt, D.; Lundborg, P.; Nyberg, G.; Raner, S.; Rönn, O.: Acute diuretic natriuretic properties of felodipine in man. Drugs *29:* suppl. 2, pp. 176–184 (1985).

70 DiBona, G.F.; Sawin, L.L.: Renal tubular site of action of felodipine. J. Pharmac. exp. Ther. *228:* 420–424 (1984).

71 Hallin, L.; Andrén, L.; Hansson, L.: Controlled trial of nifedipine and bendroflumethiazide in hypertension. J. cardiovasc. Pharmacol. *5:* 1083–1085 (1983).

72 Eggertsen, R.; Hansson, L.: Effects of treatment with nifedipine and metoprolol in essential hypertension. Eur. J. clin. Pharmacol. *21:* 389–390 (1982).

73 Opie, L.H.; Jee, L.; White, D.: Antihypertensive effects of nifedipine combined with cardioselective beta-adrenergic receptor antagonism by atenolol. Am. Heart J. *104:* 606–612 (1982).

74 Öhman, P.K.; Weiner, L.; von Schenck, H.; Karlberg, B.E.: Antihypertensive and metabolic effects of nifedipine and labetalol alone and in combination in primary hypertension. Eur. J. clin. Pharmacol. *29:* 149–154 (1985).

75 Lewis, G.R.J.; Morley, K.D.; Maslowski, A.H.; Bones, P.J.: Verapamil in the management of hypertensive patients. Aust. N.Z.J. Med. *9:* 62–64 (1979).

76 Leonetti, G.; Pasotti, C.; Ferrari, G.P.; Zanchetti, A.: Double-blind comparison of the antihypertensive effects of verapamil and propranolol; in Zanchetti, Krikler, Calcium antagonism in cardiovascular therapy: experience with verapamil, pp. 260–267 (Excerpta Medica, Amsterdam 1981).

77 Frishman, W.H.; Klein, N.A.; Klein, P.; Strom, J.A.; Tawil, R.; Strair, R.; Wong, B.; Roth, S.; LeJemtel, T.; Pollack, S.; Sonnenblick, E.H.: A comparison of oral propranolol and verapamil in patients with hypertension and angina pectoris. A placebo-controlled double-blind randomized cross-over trial. Am. J. Cardiol. *50:* 1164–1172 (1982).

78 Doyle, A.E.; Anavekar, S.N.; Oliver, L.E.: A clinical trial of verapamil in the treatment of hypertension; in Zanchetti, Krikler, Calcium antagonism in cardiovascular therapy: experience with verapamil, pp. 252–258 (Excerpta Medica, Amsterdam 1981).

79 Leary, W.P.; Phil, D.; Asmal, A.C.: Treatment of hypertension with verapamil. Curr. ther. Res. *25:* 747–752 (1979).

80 Lewis, G.R.J.; Stewart, D.J.; Lewis, B.M.; Bones, P.J.; Morley, K.D.; Janus, E.D.: The antihypertensive effect of oral verapamil – acute and long-term administration and its effects on the high-density lipoprotein values in plasma; in Zanchetti, Krikler, Calcium antagonism in cardiovascular therapy: experience with verapamil, pp. 270–277 (Excerpta Medica, Amsterdam 1981).

81 Muiesan, G.; Agabiti-Rosei, E.; Alicandri, C.; Beshi, M.; Castellano, M.; Corea, L.; Fariello, R.; Romanelli, G.; Pasini, C.; Platto, L.: Influence of verapamil on

catecholamines, renin and aldosterone in essential hypertensive patients; in Zanchetti, Krikler, Calcium antagonism and cardiovascular therapy: experience with verapamil, pp. 238–249 (Excerpta Medica, Amsterdam 1981).
82 Szalchcic, J.; Hirsch, A.T.; Tubau, J.F.; Vollmer, C.; Henderson, S.; Massie, B.M.: Diltiazem versus propranolol in essential hypertension: responses of rest and exercise blood pressure and effects on exercise capacity. Am. J. Cardiol. *59:* 393–399 (1987).
83 Schulte, K.-L.; Meyer-Sabellek, W.A.; Haertenberger, A.; Thiede, H.M.; Distler, A.; Gotzen, R.: Effects of diltiazem and nifedipine on blood pressure, plasma catecholamines and serum lipoproteins in essential hypertension. J. Hypertens. *3:* 423–424 (1985).
84 Elliot, M.L.; Pasakisi, F.; Meredith, P.A.; Reid, J.L.: Interaction between verapamil and prazosin (Abstract 475). Proc. Int. Symp. Cardiovasc. Pharmacol., Geneva 1985.
85 Hansson, L.; Dahlöf, B.; Gudbrandsson, T.; Hellsing, T.; Kullman, S.; Kuylenstierna, J.; Leppert, J.; Möller, B.; Skogström, K.; Svensson, A.; Svensson, O.; Ugander, L.: Antihypertensive effect of felodipine or hydralazine when added to beta-blocker therapy. J. cardiovasc. Pharmacol. *12:* 94–101 (1988).
86 Andersson, O.; Bengtsson, C.; Elmfeldt, D.; Haglund, K.; Hedner, T.; Saideman, P.; Strömgren, E.; Westerling, S.; Åberg, H.; Östman, J.: Short-term effects of felodipine – a new dihydropyridine – in hypertension. Br. J. clin. Pharmacol. *17:* 257–263 (1984).
87 Elmfeldt, D.; Hedner, T.: Felodipine – a new vasodilator in addition to beta-receptor blockade in hypertension. Eur. J. clin. Pharmacol. *25:* 571–575 (1983).
88 Hansson, L.; Dahlöf, B.: Antihypertensive effect of a new dihydropyridine calcium antagonist, PN 200–110 (isradipine), combined with pindolol. Am. J. Cardiol. *59:* 137B–140B (1987).
89 Orö, L.; Ryman, T.: Combination of nitrendipine with metoprolol and a comparison with nifedipine in patients insufficiently treated with the beta-blocker alone. J. cardiovasc. Pharmacol. 9 (Suppl. 4): S238–S244 (1987).
90 Guazzi, M.D.; De Cesare, N.; Galli, C.; Salvioni, A.; Tramontana, C.; Tamborini, G.; Bartorelli, A.: Calcium-channel blockade with nifedipine and angiotensin converting-enzyme inhibition with captopril in the therapy of patients with severe primary hypertension. Circulation *70:* 279–284 (1984).
91 Müller, F.B.; Bolli, P.; Erne, P.; Block, L.H.; Kiowski, W.; Bühler, F.R.: Antihypertensive therapy with the long-acting calcium antagonist nitrendipine. J. cardiovasc. Pharmacol. *6:* suppl. 7, pp. S1073–S1076 (1984).
92 Bertel, O.; Conen, D.; Radü, E.W.; Müller, J.; Lang, C.; Dubach, U.C.: Nifedipine in hypertensive emergencies. Br. med. J. *286:* 19–21 (1983).
93 Hansson, L.: Assessment of the patient's response. J. Hypertens. *3:* suppl. 2, pp. S65–S69 (1985).
94 Kazda, S.; Garthoff, B.; Luckhaus, G.: Calcium antagonists in hypertensive disease: experimental evidence for a new therapeutic concept. Post-grad. med. J. *59:* suppl. 1, pp. 78–83 (1983).
95 Lundin, S.A.; Hallbäck-Nordlander, M.I.L.: Regression of structural cardiovascular changes by antihypertensive therapy in spontaneously hypertensive rats. J. Hypertens. *2:* 11–18 (1984).
96 Weiss, R.J.; Bent, B.: Diltiazem-induced left ventricular mass regression in hypertensive patients (Abstract 415). Proc. 10th Wld. Congr. Cardiol., Washington 1986.
97 Nordlander, M.; Friberg, P.: Structural autoregulation in spontaneously hypertensive rats. Progr. appl. Microcirculat. *8:* 96–104 (1985).

98 Nyborg, N.C.B.; Mulvany, M.J.: Lack of effect of antihypertensive treatment with felodipine on cardiovascular structure of young spontaneously hypertensive rats. Cardiovasc. Res. *19:* 528–536 (1985).

99 Friberg, P.; Wåhlander, H.; Nordlander, M.: Structural and functional adaptations within the myocardium and coronary vessels after antihypertensive therapy in spontaneously hypertensive rats. J. Hypertens. *4:* suppl. 3, pp. S519–S521 (1986).

100 Henry, P.D.; Bentley, K.I.: Suppression of atherogenesis in cholesterol-fed rabbits treated with nifedipine. J. clin. Invest. *68:* 1366–1369 (1981).

101 Fleckenstein, A.; Frey, M.; Wizlebew, M. von: Vascular calcium overload – a pathogenetic factor in arteriosclerosis and its neutralization by calcium antagonists; in Kaltenbach, Neufeld, 5th Int. Adalat Symp. New Therapy for Ischemic Heart Disease and Hypertension, pp. 36–52 (Excerpta Medica, Amsterdam 1983).

102 Fleckenstein, A.: Calcium antagonists in heart and smooth muscle: experimental facts and therapeutic prospects (Whiley Interscience, New York 1983).

103 Weinstein, D.B.; Heider, J.G.: Antiatherogenic properties of calcium antagonists. Am. J. Cardiol. *59:* 163B-172B (1987).

104 Habib, J.B.; Bosaller, C.; Wells, S.; Williams, C.; Morriset, J.D.; Henry, P.D.: Preservation of endothelium-dependent vascular relaxation in cholesterol-fed rabbits by treatment with the calcium blocker PN 200–110. Circulat. Res. *58:* 305–309 (1986).

105 Henry, P.O.: Comparative pharmacology of calcium antagonists: nifedipine, verapamil and diltiazem. Am. J. Cardiol. *46:* 1047–1058 (1980).

106 Stone, P.H.; Antman, E.M.; Muller, J.E.; Braunwald, E.: Calcium channel blocking agents in the treatment of cardiovascular disorders. II. Hemodynamic effects and clinical applications. Ann. intern. Med. *93:* 886–904 (1980).

107 Lewis, G.J.: Adverse reactions to calcium antagonists. Drugs *25:* 196–222 (1983).

108 Krebs, R.: Adverse reactions with calcium antagonists. Hypertension *5:* suppl. II, pp. II-125–II-129 (1983).

109 Brouwer, R.M.L.; Follath, F.; Bühler, F.R.: Review of the cardiovascular adversity of the calcium antagonist beta-blocker combination: Implications for antihypertensive therapy. J. cardiovasc. Pharmacol. *7:* suppl. 4, pp. S38-S44 (1985).

110 Ebner, F.; Donath, M.: Mode of action and efficacy of nifedipine; in Purch, Krebs, 4th int. Adalat Symp., pp. 25–34 (Excerpta Medica, Amsterdam 1980).

111 Trost, B.N.; Weidmann, P.: Effects of nitrendipine and other calcium antagonists on glucose metabolism in man. J. cardiovasc. Pharmacol. *6:* suppl. 7, pp. S986–S995 (1984).

112 Trost, B.N.; Weidmann, P.; Beretta-Piccoli, C.: Antihypertensive therapy in diabetic patients. Hypertension *7:* suppl. II, pp. II-102–II-108 (1985).

113 Vessby, B.; Abelin, J.; Finnson, M.; Hellsing, K.; Lithell, H.: Effects of nifedipine treatment on carbohydrate and lipoprotein metabolism. Curr. ther. Res. *33:* 1075–1081 (1983).

114 Wayne, W.S.; Harper, R.W.; Laufer, E.; Federman, J.; Anderson, S.T.; Pitt, A.: Adverse interaction between beta-adrenergic blocking drugs and verapamil – report of three cases. Aust. N.Z.J. Med. *12:* 285–289 (1982).

115 Belz, G.G.; Doering, W.; Mundes, R.; Anst, P.E.; Belz, G.: Effect of various calcium antagonists on blood level and renal clearance of digoxin. Circulation *64:* suppl. IV, p. 24 (1981).

116 Pedersen, K.E.; Dorph-Pedersen, A.; Hoidt, S.; Klitgaard, N.A.; Nielsen-Kudsk, F.: Digoxin-verapamil interaction. Clin. Pharmacol. Ther. *30:* 311–316 (1981).

117 Schwartz, J.B.; Keefe, D.; Kates, R.E.; Kirsten, R.; Harrison, D.C.: Acute and chronic pharmacodynamic interaction of verapamil and digoxin in atrial fibrillation. Circulation *65:* 1163–1170 (1982).
118 Canga, L.; Sterin-Borde, L.; Gimeno, A.L.: Extracellular calcium concentration and calcium blocking agents on the action of oubain on isolated rat atrium. Pharmacol. Res. Commun. *14:* 15–30 (1982).
119 Frishman, W.H.; Charlap, S.; Kimmel, B.; Goldberger, J.; Phillipides, G.; Klein, N.: Calcium-channel blockers for combined angina pectoris and systemic hypertension. Am. J. Cardiol. *57:* suppl. D, pp. 22D–29D (1986).
120 Robertson, J.I.S.: The treatment of hypertension and quality of life. J. Hypertens. *3:* suppl. 2, S89–S90 (1985).
121 Croog, S.H.; Levine, S.; Testa, M.A.; Brown, B.; Bulpitt, C.J.; Jenkins, C.D.; Klerman, G.L.; Williams, G.H.: The effects of antihypertensive therapy on the quality of life. New. Engl. J. Med. *314:* 1657–1664 (1986).
122 Hansson, L.; Robertson, J.I.S.: Is hypertension treated adequately? Drugs 34 (Suppl. 3): 1–6 (1987).
123 Hof, R.P.: Analysis of peripheral vascular actions of the new calcium antagonist isradipine. Am. J. Med. 84 (Suppl. 3B): 18–25 (1988).

Lennart Hansson, MD, PhD, Department of Medicine, University of Göteborg, Oestra Hospital, S-41685 Göteborg (Sweden)

Van Zwieten PA (ed): Clinical Aspects of Calcium Entry Blockers.
Prog Basic Clin Pharmacol. Basel, Karger, 1989, vol 2, pp 125–142

Clinical Aspects of Calcium Entry Blockers in Cardiomyopathy, Primary Pulmonary Hypertension and Raynaud's Phenomenon

Michael P. Frenneaux[a], William J. McKenna[a], Dennis M. Krikler[b]

[a]Cardiological Sciences, St. George's Hospital Medical School, and [b]Cardiovascular Disease Unit, Royal Postgraduate Medical School, London, UK

The effects of calcium antagonists (CAs) on myocardial function and the systemic and pulmonary vasculature are examined in relation to hypertrophic cardiomyopathy, Raynaud's phenomenon, primary pulmonary hypertension and heart failure.

Hypertrophic Cardiomyopathy

Hypertrophic cardiomyopathy is a familial heart muscle disorder in which hypertrophy of the heart occurs in the absence of an apparent cause [1]. Approximately 50% of patients with hypertrophic cardiomyopathy experience symptoms of dyspnoea, chest pain or syncope [2]. The mechanisms of these symptoms are not clearly established. It has been assumed that dyspnoea is caused by rapid rise in left atrial pressure as a consequence of reduced time for filling of a poorly compliant ventricle on exercise. In general, however, it may be said that most patients with hypertrophic cardiomyopathy do not have severe limitation of exercise tolerance and the major clinical problem is unexpected sudden death which occurs with an annual mortality of 2–3% in adults and 6% in children.

The pathogenesis of hypertrophic cardiomyopathy is uncertain. Studies on the hereditary cardiomyopathy of Syrian hamsters have shown that myocardial calcium uptake and content are increased and that the metabolic and anatomical abnormalities can be prevented by the administration of verapamil [3, 4]. The relevance of this experimental finding is unclear as the cardiomyopathy of the Syrian hamster differs in many ways from hypertrophic cardiomyopathy. In the mid 1970s German workers suggested that many of the clinical features of hypertrophic cardiomyopathy could be explained by increased myocardial calcium content and they proposed that CAs may be beneficial [5, 6]. Krikler and Goodwin [7] also proposed the use of verapamil in hypertrophic cardiomyopathy. They suggested that the underlying defect in hypertrophic cardiomyopathy may include an abnormality at the cell membrane level, whereby the slow channel which conducts calcium into the cell is abnormally active. Subsequently, support for this concept has come from Berridge [8], who demonstrated that calcium acts as an important intracellular signal for cell replication and growth.

Until the late 1970s β-blockers had been the mainstay of pharmacological treatment. These are relatively effective at relieving symptoms in the majority of patients [9] but conventional doses do not appear to prevent sudden death [10].

Experience with Verapamil

There is a relatively large literature on verapamil and to a lesser extent nifedipine; there is little information on diltiazem in hypertrophic cardiomyopathy. There seems little doubt that verapamil improves symptoms in some patients with hypertrophic cardiomyopathy. In Kaltenbach and Hopf's [11] series of 61 symptomatic patients treated with high-dose verapamil (320–720 mg/day), 85% showed improvement, 11% did not change and 1 patient worsened. The NYHA functional status for the group improved from 2.7 to 1.9 [11]. The results in the series of 78 symptomatic patients of Rosing et al. [12] were slightly less impressive. Forty-two (54%) were improved over 6–30 months follow-up, and 25 (32%) improved by at least one NYHA class. However, these patients had severe symptoms which had been refractory to other therapy. The drug was discontinued in 10% of the patients during the initial hospitalisation due to adverse effects or therapeutic failure.

The relative benefit of verapamil versus propranolol for the relief of symptoms is open to question. Rosing et al. [13] gave placebo, propranolol and verapamil to 19 severely symptomatic patients with hypertrophic cardiomyopathy in a double-blind fashion for 5 days each. There was no significant difference in treadmill exercise duration between the propranolol and verapamil phases. Seven felt 'best' during the placebo phase, 9 during the verapamil phase and only 3 during the propranolol phase. It must be emphasised that these patients were highly selected in that 15 of the 19 had previously failed propranolol therapy. No randomised double-blind comparison of propranolol and verapamil in unselected patients has been performed. It has been our experience that a number of patients whose symptoms were well controlled on propranolol therapy exhibited dramatic symptomatic deterioration when they were changed to verapamil. This underscores the importance of assessing efficacy and safety of β-blockers versus verapamil in unselected patients in a controlled study.

There is objective evidence that verapamil improves exercise capacity in patients with hypertrophic cardiomyopathy. All studies have used treadmill exercise duration as the index of exercise capacity [12–16]. The improvements have generally been relatively small. Short-term verapamil administration consistently results in a small increase in treadmill exercise duration. Long-term administration of verapamil also appears to improve exercise capacity; Rosing et al. [12] reported a 53% improvement in treadmill exercise duration after 5 days therapy and a further 25% improvement at 6–30 months. There are a number of criticisms regarding the literature on verapamil's effect on exercise capacity: none of the long-term studies are placebo controlled and a training effect is therefore difficult to exclude. In addition, treadmill exercise duration is not a reproducible index of exercise capacity, and depends greatly on motivation and exercise technique. Analysis of respiratory gas exchange with measurement of maximal oxygen consumption and anaerobic threshold has been shown to be a reproducible measure of exercise capacity in patients with heart failure [17], and this would be preferable as an objective assessment of the effect of different therapeutic interventions on exercise capacity in patients with hypertrophic cardiomyopathy.

Non-invasive studies have attempted to examine the mechanism of improvement in exercise capacity by verapamil. Following short-term administration of verapamil to patients with hypertrophic cardiomyopathy, systolic function was unchanged, but diastolic function, as assessed by echocardiographic [18] and radionuclide techniques, was improved [14, 19, 20]. In the latter studies, peak filling rate increased and time to peak filling rate decreased both at rest and on exercise, and there was a relation between the degree of change and the increase in treadmill exercise duration [14]. There was return of peak filling rate and time to peak filling rate towards baseline values at 12 months, however, in spite of continued symptomatic improvement and maintenance of prolonged exercise capacity [14].

The possibility that verapamil might cause regression of left ventricular hypertrophy is raised by the calcium overload hypothesis and the observations in the Syrian hamster cardiomyopathy model. There is, however, no convincing evidence that this is the case. In Kaltenbach and Hopf's [11] series of 61 patients treated with verapamil for 10–96 months there was a significant reduction in QRS voltages (as assessed by the Sokolow index), whereas serial ECG data were available for some of the patients prior to commencing verapamil therapy and these showed increased QRS voltages. About 20% of adult patients with hypertrophic cardiomyopathy show progressive increases in QRS voltages [21] but spontaneous reductions in voltage have been reported [21, 22], usually in association with a decrease in ventricular function with fibrous replacement of active myocardium. Such changes do not therefore necessarily represent regression of the myopathic process – indeed they may imply progression. There is no convincing echocardiographic evidence of regression of left ventricular hypertrophy with verapamil [11, 15]. Kaltenbach and Hopf's [11] study also demonstrated a reduction in left ventricular muscle mass assessed angiographically, but this is not a reliable measurement. Their observation of a reduction in cardiac size radiologically probably relates to a reduction in left atrial size.

Verapamil has been widely used for the treatment of supraventricular arrhythmia in other cardiac disorders. Suggestions that CAs may be of value in some ventricular arrhythmias have come from the observations by Cranefield et al. [23] that some may be slow channel dependent, and by Wit et al. [24] that verapamil suppresses late depolarization and triggered activation. Surprisingly, there is very little literature regarding the antiarrhythmic effect of verapamil in patients with hypertrophic cardiomyopathy. McKenna et al. [25] reported 19 patients with hypertrophic cardiomyopathy who had arrhythmias which failed to respond to conventional antiarrhythmic agents and/or propranolol. Thirteen had ventricular tachycardia, 5 established atrial fibrillation with a poorly controlled ventricular response, and one had paroxysmal atrial fibrillation. They compared the effect of 1 month of each of 2 different amiodarone regimes and 1 month of verapamil therapy. Verapamil was ineffective at treating these arrhythmias, indeed there was an increase in the number of patients with frequent and complex ventricular extrasystoles and ventricular tachycardia. Furthermore, 3 patients developed pulmonary oedema following commencement of verapamil.

The question of whether verapamil reduces the risk of sudden death has not been answered. No large-scale randomised placebo-controlled trial on an unselected group of patients with hypertrophic cardiomyopathy has been performed. In Kaltenbach and Hopf's [11] series, the mortality rate in patients treated with verapamil was 1.3% per patient year. The mortality rate in a group of adult patients with hypertrophic cardiomyopathy is usually of the order of 2–3%, but their series is selected and uncontrolled.

Given the high dosages of verapamil used in the treatment of hypertrophic cardiomyopathy (up to 760 mg/day), it is not surprising that adverse effects such as headache, vertigo gastro-intestinal disturbances, peripheral oedema, postural hypotension and prolongation of the P–R interval were common [26]. Though troublesome, such side effects have not usually required withdrawal of the drug. More serious adverse effects have, however, been reported. They were rare in Kaltenbach and Hopf's [11] series of 61 patients but more common in Epstein and Rosing's [26] series of 120 patients: (1) sinus bradycardia with junctional escape rhythm or isorhythmic dissociation (11%); (2) Mobitz type I 2nd degree A-V block (4 patients); (3) Mobitz type II A-V block (1 patient); (4) profound hypotensive collapse (2 patients), and (5) pulmonary oedema. In Epstein and Rosing's series this developed in 8 of 120 patients, of whom 3 died. Of 19 patients given verapamil by McKenna et al. [25], 3 developed pulmonary oedema. Various mechanisms may be responsible. Vasodilation may result in an increase in obstruction or impaired filling with consequent decreased cardiac output leading to reduced coronary perfusion and deleterious effects on cardiac function. In the small subset of patients with poor systolic function, the negative inotropic effect of verapamil may be of some importance. Complete heart block has also been reported and may result in sudden death [27].

Reports of the use of verapamil in children with hypertrophic cardiomyopathy are sparse. Spicer et al. [28] reported the effect of verapamil given to 13 'children' with hypertrophic cardiomyopathy (age 7 months to 19 years, mean 12 years) for an average of 13 months. Symptoms improved. There was a small (33%) improvement in exercise duration; this could be explained on the basis of a training effect and the fact that the children were older at the second test. QRS voltages did not change, but when adjusted for body size there was a significant reduction in left ventricular septal and posterior wall thickness.

Invasive Haemodynamic Studies

Most acute haemodynamic studies with verapamil have demonstrated falls in systolic blood pressure, systemic vascular resistance and gradients and a rise in cardiac index [15, 29]. The effect on heart rate is variable. Some studies have demonstrated no change in heart rate [15] and others an increase [29]. This presumably depends on the balance between the direct suppressant effect on sinus node function and the effect of increased sympathetic tone mediated by the baroreceptors as a result of hypotension. In view of the symptomatic improvement and the radionuclide and echocardiographic evidence of improvement in indices of diastolic function, one might have expected infusions of verapamil to lower pulmonary capillary wedge pressure and left-ventricular end-diastolic pressure but numerous studies have revealed no significant change in either [15, 29]. Indeed the clinical observation that verapamil can precipitate pulmonary oedema in a small proportion of patients implies that the converse may occur [25, 26]. The situation is further complicated by the observations of Wilmshurst et al. [30]. Using a Millar catheter, they assessed systolic and diastolic function at various heart rates (achieved by atrial pacing) before and after the acute administration of verapamil. Verapamil did not change diastolic indices, but reduced systolic indices. They also observed (by coronary sinus sampling) that verapamil did not have any consistent effect on anaerobic myocardial metabolism. They suggested that the beneficial symptomatic effect of verapamil related to a reduction in unproductive contraction and hence myocardial oxygen consumption and wall stress. This concept requires further evaluation.

The effect of long-term verapamil therapy on haemodynamics has been investigated in a number of studies and the results are conflicting. Kaltenbach and Hopf [11] reported a reduction in gradients but no change in left-ventricular systolic and diastolic volumes, pulmonary capillary wedge pressure or left-ventricular end-diastolic pressure, Chatterjee et al. [31] confirmed the lack of change in left-ventricular end-diastolic pressure but demonstrated an increase in stroke volume and left-ventricular end-diastolic volume.

Thus no clear explanation for the beneficial symptomatic effect of verapamil is forthcoming from the available haemodynamic data. No studies have looked at the effect of verapamil on exercise haemodynamics – and it may be that this would provide more information.

Experience with Nifedipine

Suggestions for the use of nifedipine in hypertrophic cardiomyopathy arose from the case of a 54-year-old lady with severe hypertrophic cardiomyopathy reported by Lorell et al. [32]. Administration of nifedipine resulted in marked symptomatic improvement in association with reduction in isovolumic relaxation time, normalisation of the previously abnormal left-ventricular pressure and dimension. Nifedipine, unlike β-blockers, verapamil and amiodarone, does not slow heart rate and any beneficial effect must relate either to a reduction in afterload or a direct effect on the heart muscle resulting in improvement in diastolic compliance. In a subsequent study of 15 patients with hypertrophic cardiomyopathy, Lorell et al. [33] demonstrated a reduction in isovolumic relaxation time and an increase in the rate of left-ventricular filling and posterior wall thinning. They subsequently asserted that this improvement in diastolic function could not be accounted for solely on the basis of afterload reduction, because nitroprusside titrated to produce a similar fall in systolic blood pressure resulted in a considerably lesser improvement in ventricular relaxation and filling indices [34]. Betocchi et al. [35], however, failed to demonstrate any improvement in diastolic function with nifedipine using simultaneous gated radionuclide angiography and Millar catheter haemodynamic measurements.

The hopes raised by the elegant studies of Lorell et al. [33] have not been met in clinical practice. Nifedipine cannot be recommended as a first- or second-line treatment for patients with hypertrophic cardiomyopathy. The literature consists mainly of case reports and small uncontrolled studies, but the consensus is that nifedipine is inferior to either the β-blocking agents [36] or verapamil [37] and in a series of 11 patients treated with nifedipine Ruddy et al. [38] reported the development of pulmonary oedema in 2.

Raynaud's Phenomenon

Raynaud's phenomenon may be either primary (Raynaud's disease) or secondary to a variety of other disorders such as scleroderma, mixed connective tissue disease, Buerger's disease and exposure to vibrating tools. Primary Raynaud's phenomenon probably accounts for about 90% of those seen in general practice [39], but the secondary forms are disproportionately common in hospital practice. Heslop et al. [40], in a large-scale study, reported that Raynaud's phenomenon was more common than had been appreciated – occurring in 5% of men and 10% of women, but in the majority of cases was not perceived by the patient as a serious problem. Nevertheless for some patients it is a major cause of morbidity.

The concept that Raynaud's phenomenon is a manifestation of a more generalised vasospastic disorder has arisen from its reported association with migraine [41] and variant angina [42] and from the demonstration that cold-induced pulmonary vasospasm occurs in patients with Raynaud's disease [43].

In some patients simple measures involving avoidance of exposure to cold are effective. Sympathectomy and various pharmacological agents including sublingual and topical nitrates [44–46], prostaglandins [47, 48], angiotensin-converting enzyme inhibitors [49], α-adrenergic blocking agents [50] and thromboxane synthetase inhibitors [51] have been used with differing degrees of success. None of these can be recommended as the definitive treatment.

The first report of the use of CAs in the treatment of Raynaud's phenomenon was from Kahan et al. [52] in 1981. They found that a single oral dose of 20 mg nifedipine delayed the onset of Raynaud's phenomenon during a cold challenge and demonstrated long-term symptomatic improvement from the drug in an uncontrolled study.

Subsequently, a number of studies have evaluated the effectiveness of CAs in Raynaud's phenomenon, in the majority of these, nifedipine was used. There seems little doubt that nifedipine produces symptomatic improvement [53–59], particularly in the primary form [60], but the results of objective assessment of digital blood flow are less clear-cut. Whereas Winston et al. [54] have reported a small reduction in finger tip vascular resistance in 5 of 6 patients given nifedipine and Nilsson et al. [62] have shown increased resistance to cold challenge as assessed by digital BP and temperature after nifedipine, other studies have failed to demonstrate an objective improvement in digital blood flow [55, 59, 61].

Primary Pulmonary Hypertension

Primary pulmonary hypertension was first described by Wood [63, 64]. It is more common in women of childbearing age and symptoms often develop in relation to puberty, childbirth and oral contraceptive therapy [65, 66]. It has been linked to two anoretic agents, aminorex fumarate (no longer available) and fenfluramine (still available). It is more common in populations living at high altitude [67] and there is an association with Raynaud's phenomenon [68]. Familial cases have been reported.

In Wood's original series, the median survival was only 2 years [63, 64] and other studies have reported similarly poor survival rates [69, 70]. A recent report from the Hammersmith group [71], however, suggested that the outlook may not be quite as gloomy. Twelve of thirty-four patients in their series survived for more than 5 years. Clinical evidence of right heart failure, ECG evidence of right-ventricular strain, right-ventricular enlargement on chest X-ray and low cardiac output at presentation are markers of a poor prognosis [71, 72].

The concept that the underlying disease process may be reversible was originally raised by Wood [73]. He demonstrated that pulmonary vascular resistance could be selectively lowered in some but not all patients with primary pulmonary hypertension by infusion of acetylcholine, whereas patients with Eisenmenger syndrome were universally unresponsive.

A spectrum of histological change has been demonstrated. The earliest feature is medial hypertrophy of the pulmonary arterioles and this progresses through intimal proliferation, obliterative fibrosis and fibrinoid necrosis, to the most severe change, 'plexiform arteriopathy' of Wagenvoort [74]. The current concept is that the milder histological changes may be associated with marked reactivity of the pulmonary vascular resistance, whereas the more advanced changes are associated with a predominantly fixed increase in pulmonary vascular resistance. Although vasodilators may reverse the reactive component of the increased pulmonary vascular resistance it remains to be established whether this may halt progression of the underlying disease process and influence survival. The ideal vasodilator agent would be suitable for long-term use and would preferentially vasodilate the pulmonary rather than the systemic vascular bed. No such agent currently exists. In patients with right heart failure and a relatively fixed cardiac output, marked reduction in systemic vascular resistance and reduction in venous return may result in profound haemodynamic collapse. Various vasodilator drugs have been assessed, including prostacyclin infusions, hydralazine, phentolamine diazoxide and nitroglycerine. The results have been mixed [75–77]. None of these agents has been shown to improve survival.

The calcium channel blockers, particularly nifedipine, have also been used in the treatment of primary pulmonary hypertension. In 1980, Camerini et al. [78] reported a 34-year-old lady with severe primary pulmonary hypertension in whom sublingual nifedipine resulted in a 54%

fall in pulmonary vascular resistance and a 14% decrease in pulmonary artery pressure. Long-term treatment of this patient with nifedipine resulted in symptomatic improvement and a reduction in the signs of right heart failure. There are a relatively larger number of case reports in the literature documenting short- and long-term symptomatic and haemodynamic improvements with nifedipine, verapamil and diltiazem. There are very few larger series and no placebo-controlled trials.

Rubin et al. [79] demonstrated a significant increase in cardiac output and reduction in pulmonary vascular resistance without significant change in pulmonary artery pressure in patients given sublingual nifedipine acutely. Right-ventricular volumes and ejection fraction (assessed by radionuclide angiography) improved. Six of these patients were placed on long-term nifedipine therapy and persistent haemodynamic and symptomatic improvement was seen in 5. Olvari reported long-term symptomatic improvement and increased treadmill exercise duration in a group of 7 patients given nifedipine [80], and a similar improvement was documented by Hughes et al. [81] in 4 of 6 patients. Conversely, other reports have documented haemodynamic deterioration after sublingual nifedipine [82, 83]. In one patient with a persistent foramen ovale, nifedipine resulted in massive right to left shunting as a result of preferential systemic vasodilatation [84]. Patients with severe primary pulmonary hypertension often have a failing right ventricle and in these the negative inotropic effect of CAs may be detrimental.

The absence of large-scale placebo-controlled studies makes it impossible to say whether the isolated reports described above of symptomatic improvement with the CAs represent anything more than chance variations in the natural history of the disease. Until such trials are published, it seems reasonable to assess the acute haemodynamic effect of vasodilatation including CAs in the individual patient and to continue such therapy long term in those who have shown acute benefit. Heart-lung transplantation currently offers the best hope of symptomatic improvement and perhaps of prolonged survival for the patient with severe and rapidly progressive symptoms and poor prognostic markers.

Cardiac Failure

In patients with heart failure, drugs which reduce afterload increase cardiac output and reduce myocardial oxygen consumption [85, 86]. The vasodilator action of the calcium channel antagonists [87] has led to the suggestion that these agents may be useful in the treatment of heart failure. In addition their direct coronary vasodilator action and the observation that they stabilise cell membranes and prevent progressive myocardial cell death [88] may have particular relevance when the underlying basis for the heart failure is ischaemic heart disease.

In practice, however, the calcium channel blockers also have significant negative inotropic actions which tend to offset the above beneficial effects. Of the first-generation drugs, the dihydropyridine derivative nifedipine exhibits the least negative inotropic effects and early attention therefore focussed principally on this agent for the treatment of heart failure. A number of studies have demonstrated beneficial haemodynamic effects following the acute administration of sublingual nifedipine – with reductions in aortic pressure, systemic vascular resistance and pulmonary artery pressure and increases in ejection fraction and cardiac index [89–92]. The effect on left- and right-sided filling pressures was, however, inconsistent [89, 92]. Haemodynamic improvement persists for at least 6 h following a single sublingual dose of nifedipine [92] is maintained following 2 months of oral nifedipine therapy, and is lost within 48 h of withdrawal of the drug [93]. However, there are also reports of nifedipine precipitating pulmonary oedema in patients with heart failure [94, 95]. Though haemodynamic studies are encouraging the literature is lacking in controlled trials, particularly with regard to symptoms and objective evidence of effects on exercise capacity, and in general clinical experience with nifedipine has been disappointing in patients with heart failure.

The second generation dihydropyridine agents nicardipine and felodipine have negligible negative inotropic effects [96], whilst retaining potent vasodilator activity and this represents a major advance. Acute haemodynamic studies performed at rest and on exercise with nicardipine have demonstrated reductions in arterial pressure, systemic vascular resistance, ejection time, pulmonary capillary wedge pressure with reduced arteriovenous oxygen difference at peak exercise and increased heart rate, cardiac output, ejection fraction and ejection rate [97–99]. The improvements in cardiac output, ejection fraction and ejection rate persist for at least 30 min after discontinuing a 10-min intravenous infusion [97]. These improvements in both resting and exercise haemodynamics are maintained after 1 week of oral nicardipine therapy but no improvement in maximal oxygen consumption could be detected [100]. The latter observation is not surprising, however, since therapeutic interventions in heart failure are known to take a considerable time to produce objective increases in exercise capacity even though haemodynamic improvement may occur very early. This is thought to be due, at least in part, to a delay in resolution of factors which impair skeletal muscle blood flow [101]. In a 6-week uncontrolled study of nicardipine in 15 patients with NYHA Classes II and III heart failure, symptoms and exercise duration improved and weight fell in 13 [102]. Although encouraging, these results must be interpreted with caution, and await confirmation in a placebo controlled cross-over study with objective assessment of exercise capacity by respiratory gas analysis. Though such long-term studies are lacking activation of the renin-angiotensin system with later attenuation of the beneficial early effects as occurs with other non ACE inhibitor vasodilators might be anticipated.

References

1 Report of the WHO/ISFC task force on the definition and classification of cardiomyopathies. Br. Heart J. *44:* 672–673 (1980).

2 Maron, B.J.; Bonow, R.O.; Cannon, R.O.III; Leon, M.B.; Epstein, S.E.: Hypertrophic cardiomyopathy. Interrelations of clinical manifestations, pathophysiology, and therapy. New Engl. J. Med. *316:* 780–789 (1987).

3 Bajusz, E.; Lossnitzer, K.: Ein neues Krankheitsmodell: Erbliche nicht vaskulare Myokarddegeneration mit Herzinsuffizienz (Letter). Münch. med. Wschr. *110:* 1756–1768 (1968).

4 Lossnitzer, K.: Genetic induction of cardiomyopathy; in Schnuer, Erlicher, Heart and circulation, Handbuch der experimentellen Pharmakologie, vol. 13, p. 309 (-010, 1975).

5 Fleckenstein, A.: Specific pharmacology of calcium in myocardium, cardiac pacemakers, and vascular smooth muscle. Annu. Rev. Pharmacol. Toxicol. *17:* 149–166 (1977).

6 Kaltenbach, M.; Hopf, R.; Kelle, M.: Calcium-antagonistische Therapie bei hypertroph-obstruktiver Kardiomyopathie. Dt. med. Wschr. *101:* 1284–1287 (1976).

7 Goodwin, J.F.; Krikler, D.M.: Arrhythmia as a cause of sudden death in hypertrophic cardiomyopathy. Lancet *ii:* 937–939 (1976).

8 Berridge, M.J.: The molecular basis of communication within the cell. Scient. Am. *253:* 142–152 (1985).

9 Cherian, G.; Brockington, I.F.; Shah, P.M.; Oakley, C.M.; Goodwin, J.F.: Beta-adrenergic blockade in hypertrophic obstructive cardiomyopathy. Br. med. J. *i:* 895–898 (1966).

10 McKenna, W.J.; Goodwin, J.F.: The natural history of hypertrophic cardiomyopathy. Curr. Probl. Cardiol. *VI:* 5–26 (1981).

11 Kaltenbach, M.; Hopf, R.: Treatment of hypertrophic cardiomyopathy: relation to pathological mechanisms. J. mol. Cell. Cardiol. *17:* suppl. 2, pp. 59–68 (1985).

12 Rosing, D.R.; Condit, J.R.; Maron, B.J.; Kent, K.M.; Leon, M.B.; Bonow, R.O.; Lipson, L.C.; Epstein, S.E.: Verapamil therapy: a new approach to the pharmacologic

treatment of hypertrophic cardiomyopathy. III. Effects of long-term administration. Am. J. Cardiol. *48:* 545–553 (1981).

13 Rosing, D.R.; Kent, K.M.; Maron, B.J.; Epstein, S.E.: Verapamil therapy: a new approach to the pharmacologic treatment of hypertrophic cardiomyopathy. II. Effects on exercise capacity and symptomatic status. Circulation *60:* 1208–1213 (1979).

14 Bonow, R.O.; Dilsizian, V.; Rosing, D.R.; Maron, B.J.; Bacharach, S.L.; Green, M.V.: Verapamil induced improvement in left ventricular filling and increased exercise tolerance in patients with hypertrophic cardiomyopathy: short- and long-term effects. Circulation *72:* 853–864 (1985).

15 Rosing, D.R.; Idanpaan-Heikkla, U.; Maron, B.J.; Bonow, R.O.; Epstein, S.E.: Use of calcium channel blocking drugs in hypertrophic cardiomyopathy. Am. J. Cardiol. *55:* 185B–195B (1985).

16 Rosing, D.R.; Cannon, R.O.III; Watson, R.M.; Kent, K.M.; Lakatos, E.; Epstein, S.E.: Comparison of verapamil and nifedipine. Effects on symptoms and exercise capacity in patients with hypertrophic cardiomyopathy. Circulation *66:* suppl. II, pp. II–24 (1982).

17 Lipkin, D.P.; Perrins, J.; Poole-Wilson, P.A.: Respiratory gas exchange in the assessment of patients with impaired left ventricular function. Br. Heart J. *54:* 321–328 (1985).

18 Hanrath, P.; Mathey, D.G.; Kremer, P.; Sonntag, F.; Bleifeld, W.: Effect of verapamil on left ventricular isovolumic relaxation time and regional left ventricular filling in hypertrophic cardiomyopathy. Am. J. Cardiol. *45:* 1258–1264 (1980).

19 Bonow, R.O.; Rosing, D.R.; Bacharach, S.L.; Green, M.V.; Kent, K.M.; Lipson, L.C.; Maron, B.J.; Leon, M.B.; Epstein, S.E.: Effects of verapamil on left ventricular systolic function and diastolic filling in patients with hypertrophic cardiomyopathy. Circulation *64:* 787–796 (1981).

20 Bonow, R.O.; Frederick, T.M.; Bacharach, S.L.; Green, M.V.; Goose, P.W.; Rosing, D.R.: Atrial systole and left ventricular filling in hypertrophic cardiomyopathy: effect of verapamil. Am. J. Cardiol. *51:* 1386–1391 (1983).

21 McKenna, W.J.; Borggrefe, M.; England, D.; Deanfield, J.; Oakley, C.M.; Goodwin, J.F.: The natural history of left ventricular hypertrophy in hypertrophic cardiomyopathy: an electrocardiographic study. Circulation *66:* 1233–1240 (1982).

22 Spirito, P.; Maron, B.J.; Bonow, R.O.; Epstein, S.E.: Severe functional limitation in patients with hypertrophic cardiomyopathy and only mild localized left ventricular hypertrophy. J. Am. Coll. Cardiol. *8:* 537–544 (1986).

23 Cranefield, P.F.; Aronson, R.S.; Wit, A.L.: Effect of verapamil on the normal action potential and on a calcium-dependent slow response of canine Purkinje fibers. Circulation Res. *34:* 204–213 (1974).

24 Wit, A.; Boyden, P.A.; Gadsby, D.C.; Cranefield, P.F.: Triggered activity as a cause of atrial arrhythmias; in Narula, Cardiac arrhythmias: electrophysiology, diagnosis and management, pp. 14–31 (Williams & Wilkins, Baltimore 1979).

25 McKenna, W.J.; Harris, L.; Perez, G.; Krikler, D.M.; Oakley, C.M.; Goodwin, J.F.: Arrhythmia in hypertrophic cardiomyopathy. II. Comparison of amiodarone and verapamil in treatment. Br. Heart J. *46:* 173–178 (1981).

26 Epstein, S.E.; Rosing, D.R.: Verapamil: its potential for causing serious complications in patients with hypertrophic cardiomyopathy. Circulation *64:* 437–441 (1981).

27 Perrot, B.; Danchin, D.; Terrier de la Chause, A.: Verapamil: a cause of sudden death in a patient with hypertrophic cardiomyopathy. Br. Heart J. *51:* 352–354 (1984).
28 Spicer, R.L.; Rocchini, A.P.; Crowley, D.C.; Rosenthal, A.: Chronic verapamil therapy in pediatric and young adult patients with hypertrophic cardiomyopathy. Am. J. Cardiol. *53:* 1614–1619 (1984).
29 Rosing, D.R.; Kent, K.M.; Borer, J.S.; Seides, S.F.; Maron, B.J.; Epstein, S.E.: Verapamil therapy. A new approach to the pharmacological treatment of hypertrophic cardiomyopathy. I. Hemodynamic effects. Circulation *60:* 1201–1207 (1979).
30 Wilmshurst, P.T.; Thompson, D.S.; Juul, S.M.; Jenkins, B.S.; Webb-Peploe, M.M.: Effects of verapamil on haemodynamic function and myocardial metabolism in patients with hypertrophic cardiomyopathy. Br. Heart, J. *56:* 544–553 (1986).
31 Chatterjee, K.; Roff, G.; Anderson, D.; et al.: Effects of chronic verapamil therapy on systolic and diastolic function in hypertrophic cardiomyopathy: angiographic study (Abstract). Proc. of Diastolic Function of the Heart, Int. Symp. Hamburg 1982.
32 Lorell, B.H.; Paulus, W.J.; Grossman, W.; Wynne, J.; Cohn, P.F.; Braunwald, E.: Improved diastolic function and systolic performance in hypertrophic cardiomyopathy after nifedipine. New Engl. J. Med. *303:* 801–803 (1980).
33 Lorell, B.H.; Paulus, W.J.; Grossmann, W.; Wynne, J.; Cohn, P.F.: Modification of abnormal left ventricular diastolic properties by nifedipine in patients with hypertrophic cardiomyopathy. Circulation *65:* 499–507 (1982).
34 Paulus, W.J.; Lorell, B.H.; Craig, W.E.; Wynn, J.; Murgo, J.P.; Grossmann, W.: Comparison of the effects of nitroprusside and nifedipine on diastolic properties in patients with hypertrophic cardiomyopathy: altered left ventricular loading or improved muscle inactivation? J. Am. Coll. Cardiol. *5:* 879–886 (1983).
35 Betocchi, S.; Cannon, R.O. III; Watson, R.M.; Bonow, R.O.; Ostrow, H.G.; Epstein, S.E.; Rosing, D.R.: Effects of sublingual nifedipine on haemodynamics and systolic and diastolic function in patients with hypertrophic cardiomyopathy. Circulation *72:* 1001–1007 (1985).
36 Koide, T.; Kakihana, M.; Takabatake, Y.; Jizuka, M.; Uchida, Y.; Ozeki, K.; Morooka, S.; Kato, A.; Tanaka, S.; Oya, T.; Momomura, S.; Muruo, S.: Long-term clinical effect of calcium inhibitors in hypertrophic cardiomyopathy compared to the effects of beta blocking agents: a preliminary report with special reference to the beneficial effects of nifedipine on angina pectoris. Jap. Heart J. *22:* 87–102 (1981).
37 Rosing, D.R.; Cannon, R.O.III; Watson, R.M.; Kent, K.M.; Lakatos, E.; Epstein, S.E.: Comparison of verapamil and nifedipine effects on symptoms and exercise capacity in patients with hypertrophic cardiomyopathy (Abstract). Circulation *66:* suppl. II, p. 24 (1982).
38 Ruddy, T.D.; Koilpillai, C.; Liu, P.P.; Mahon, W.A.; Henderson, M.A.; McLaughlin, P.R.; Rakowski, H.; Wigle, E.D.: Evaluation of chronic nifedipine therapy in nonobstructive hypertrophic cardiomyopathy (Abstract). Circulation *66:* suppl. II, p. 24 (1982).
39 Coffman, J.D.; Davies, T.: Vasospastic disorders: a review. Prog. cardiovasc. Dis. *18:* 123–146 (1975).
40 Heslop, J.; Coggon, D.; Acheson, E.D.: The prevalence of intermittent digital ischaemia (Raynaud's phenomenon) in a general practice. J. R. Coll. gen Pract. *33:* 85–89 (1983).

41 Zahavi, I.; Chagnac, A.; Herring, R.; Davidovic, H.; Kuritzky, A.: Prevalence of Raynaud's phenomenon in patients with migraine. Archs intern. Med. *144:* 742–744 (1984).
42 Robertson, D.; Oates, J.A.: Variant angina and Raynaud's phenomenon (Letter). Lancet. *i:* 452 (1978).
43 Fahey, P.J.; Utell, M.J.; Conderin, J.J.; Green, R.; Hyde, R.: Raynaud's phenomenon of the lung. Am. J. Med. *76:* 263–269 (1984).
44 Peterson, L.; Vorhies, C.: Raynaud's syndrome, treatment with sublingual administration of nitroglycerin, swinging arm manoeuvre and biofeedback training. Archs Derm. *119:* 396–399 (1983).
45 Klechner, M.S.; Allen, E.; Khalil, E.: The effect of local application of glyceryl trinitrate (nitroglycerin) on Raynaud's disease and Raynaud's phenomenon. Circulation *3:* 681–689 (1951).
46 Diem, C.; Mueller-Buehl, U.; Morl, H.: Isosorbide dinitrate ointment in Raynaud's disease. Z. Kardiol. *72:* suppl. 3, pp. 185–187 (1983).
47 Clifford, P.C.; Martin, M.; Sheddon, E.; Kirby, J.; Baird, R.; Dieppe, P.: Treatment of vasospastic disease with prostaglandin E1. Br. med. J. *280:* 1031–1034 (1980).
48 Belch, J.; Drury, J.; Capell, H.; Forbes, C.; Newman, P.; McKenzie, F.; Lieberman, P.; Prentice, C.: Intermittent epoprostenol (prostacyclin) infusion in patients with Raynaud's phenomenon. Lancet *i:* 313–315 (1983).
49 Lopez-Ojevero, J.; Sall, S.; D'Angelo, W.: Reversal of vascular and renal crisis of scleroderma by oral angiotensin converting enzyme blockade. New Engl. J. Med. *300:* 1417–1419 (1979).
50 Waldo, R.: Prazosin relieves Raynaud's vasospasm (Letter) J. Am. med. Ass. *241:* 1037 (1979).
51 Belch, J.; Cormie, J.; Newman, P.; McLaren, M.; Barbenel, J.; Capell H.; Lieberman, P.; Forbes, C.; Prentice, C.: Dazoxiben, a thromboxane synthetase inhibitor, in the treatment of Raynaud's syndrome: a double-blind trial. Br. J. clin. Pharmacol. *15:* 1135–1165 (1983).
52 Kahan, A.; Weber, S.; Amar, B.; Saporta, L.; Hodara, M.: Nifedipine and Raynaud's phenomenon (Letter). Ann. intern. Med. *94:* 546 (1981).
53 Nilsson, H.; Jonasson, T.; Leppert, J.; Ringquist, I.: Nifedipine in cold-induced vasospasm (Abstract). Eur. Heart J. *5:* suppl. 1, p. 33 (1984).
54 Winston, E.; Pariser, K.; Miller, K.; Salem, D.; Creager, M.: Nifedipine as a therapeutic modality for Raynaud's phenomenon. Arthritis Rheum. *26:* 1177–1180 (1983).
55 Smith, C.D.; McKendry, R.: Controlled trial of nifedipine in the treatment of Raynaud's phenomenon. Lancet. *ii:* 1299–1301 (1982).
56 Nayssairat, M.; Caprou, L.; Fiessinger, J.; Mathieu, J.; Housset, E.: Calcium channel blockers and Raynaud's disease. Ann. intern. Med. *95:* 243 (1981).
57 Kahan, A.; Weber, S.; Amor, B.; Menkes, C.; Saporta, L.; Hadora, M.; Guerin, G.; Degeorges, M.: Calcium entry blocking agents in digital vasospasm (Raynaud's phenomenon). Eur. Heart J. *4:* suppl. C, pp. 123–129 (1983).
58 Sanza, J.; Kraus, A.; Gonzales-Amaro, R.; Alarcon-Segovia, D.: Effect of the calcium-channel blocker nifedipine on Raynaud's phenomenon. A controlled double-blind trial. J. Rheumatol. *11:* 362–364 (1984).

59 Aldoori, M.; Campbell, W.B.; Dieppe, P.A.: Nifedipine in the treatment of Raynaud's syndrome. Cardiovasc. Res. *20:* 466–470 (1986).

60 Rodeheffer, R.; Rommer, J.; Wigley, F.; Smith, C.: Controlled double-blind trial of nifedipine in the treatment of Raynaud's phenomenon. New Engl. J. Med. *308:* 880–883 (1983).

61 Ettinger, W.; Wise, R.; Schaffhauser, D.; Wigley, F.: Controlled double-blind trial of dazoxiben and nifedipine in the treatment of Raynaud's phenomenon. Am. J. Med. *77:* 451–458 (1984).

62 Nilsson, H.; Jonasson, T.; Ringquist, I.: Treatment of digital vasospastic disease with the calcium entry blocker nifedipine. Acta med. scand. *215:* 135–139 (1984).

63 Wood, P.: Congenital heart disease. Br. med. J. *ii:* 693–699 (1950).

64 Wood, P.: Pulmonary hypertension. Br. med. Bull. *8:* 348–353 (1952).

65 Oakley, C.M.; Somerville, J.: Oral contraceptives and progressive pulmonary vascular disease. Lancet *i:* 890–893 (1968).

66 Kleiger, R.E.; Boxer, M.; Ingham, R.; Harrison, D.C.: Pulmonary hypertension in patients using oral contraceptives. Chest *69:* 143–147 (1976).

67 Vogel, J.H.K.; Weaver, W.F.; Rose, R.L.; Blount, S.G. Jr.; Grover, R.F.: Pulmonary hypertension on exertion in normal man living at 10,150 feet (Leadville, Colorado). Med. thorac. *19:* 461–477 (1962).

68 Wade, G.; Ball, J.: Unexplained pulmonary hypertension. Q. J. Med. *26:* 83–119 (1957).

69 Walcott, G.; Burchell, H.B.; Brown, A.L.J.: Primary pulmonary hypertension. Am. J. Med. *49:* 79–79 (1970).

70 Fuster, V.; Giuliani, E.R.; Brandenburg, R.O.; Weidman, W.H.; Edwards, W.D.: The natural history of idiopathic pulmonary hypertension (Abstract). Am. J. Cardiol. *47:* 422 (1981).

71 Rozkovec, A.; Montanes, P.; Oakley, C.M.: Factors that influence the outcome of primary pulmonary hypertension. Br. Heart J. *55:* 449–458 (1986).

72 Fuster, V.; Stede, P.M.; Edwards, W.D.; Gersh, B.J.; McGoon, M.D.; Frye, R.L.: Primary pulmonary hypertension: natural history and the importance of thrombosis. Circulation *70:* 580–587 (1984).

73 Wood, P.: The vasoconstrictive factor in pulmonary hypertension; in Adams, Veith, Pulmonary circulation, pp. 294–301 (Grune & Stratton, New York 1959).

74 Wagenvoort, C.A.; Wagenvoort, N.: Primary pulmonary hypertension: a pathologic study of the lung vessels in 156 clinically diagnosed cases. Circulation *42:* 1163–1184 (1970).

75 Packer, M.; Greenberg, B.; Massie, B.; Dash, H.: Deleterious effects of hydralazine in patients with pulmonary hypertension. New Engl. J. Med. *306:* 1326–1331 (1982).

76 Buch, J.; Wennevold, A.: Hazards of diazoxide in pulmonary hypertension. Br. Heart J. *46:* 401–403 (1981).

77 Cohen, M.L.; Kronzon, I.: Adverse hemodynamic effects of phentolamine in primary pulmonary hypertension. Ann. intern. Med. *95:* 591–592 (1981).

78 Camerini, F.; Alberti, E.; Klugmann, S.; Salvi, A.: Primary pulmonary hypertension: effects of nifedipine. Br. Heart J. *44:* 352–356 (1980).

79 Rubin, L.J.; Nicod, P.; Hillis, L.D.; Firth, B.G.: Treatment of primary pulmonary hypertension with nifedipine: a hemodynamic and scintigraphic evaluation. Ann. intern. Med. *99:* 433–438 (1983).

80 Olivari, M.T.; Levine, T.B.; Weir, E.K.; Cohn, J.N.: Hemodynamic effects of

nifedipine at rest and during exercise in primary pulmonary hypertension. Chest *86:* 14–19 (1984).
81 Hughes, J. D.; Rubin, L. J.: Primary pulmonary hypertension. An analysis of 28 cases and a review of the literature. Medicine *65:* 56–72 (1986).
82 Kastanos, N.; Estopa, R.; Rodriguez-Roisen, R.; Agusti-Vidal, A.: Hydralazine in pulmonary hypertension (Letter) New Engl. J. Med. *307:* 1215 (1982).
83 Packer, M.; Greenberg, B.; Massie, B.; Dash, H.: Hydralazin in pulmonary hypertension (Letter). New Engl. J. Med. *307:* 1215–1216 (1982).
84 Laine, J. F.; Slama, M.; Petitpretz, P.; Girard, P.; Motte, G.: Danger of vasodilator therapy of pulmonary hypertension in patent foramen ovale. Chest *89:* 894–895 (1986).
85 Sonnenblick, E. H.; Downing, S. E.: Afterload as a primary determinant of ventricular performance. Am. J. Physiol. *204:* 604–610 (1963).
86 Cohn, J. N.; Franciosa, J. A.: Vasodilator therapy of cardiac failure. New Engl. J. Med. *27:* 297–331 (1977).
87 Massingham, R.: A study of compounds which inhibit vascular smooth muscle contraction. Eur. J. Pharmacol. *22:* 75–82 (1973).
88 Factor, S. M.; Minase, T.; Cho, S.; Dominitz, R.; Sonnenblick, E. H.: Microvascular spasm in the cardiomyopathic Syrian hamster: a preventable cause of focal myocardial necrosis. Circulation *66:* 342–354 (1982).
89 Klugmann, S.; Salvi, A.; Camerini, F.: Haemodynamic effects of nifedipine in heart failure. Br. Heart J. *43:* 440–446 (1980).
90 Ludbrook, P. A.; Tiefenbrunn, A. J.; Reed, F. R.; Sobel, B. E.: Acute hemodynamic responses to sublingual nifedipine: dependence on left ventricular function. Circulation *65:* 489–497 (1982).
91 Hanrath, P.; Kremer, P.; Bleifeld, W.: Influence of nifedipine on left ventricular dysfunction at rest and during exercise. Eur. Heart J. *3:* 325–330 (1982).
92 Elkayam, U.; Weber, L.; Torkan, B.; Berman, D.; Rahimtoola, S. H.: Acute hemodynamic effect of oral nifedipine in severe chronic congestive heart failure. Am. J. Cardiol. *52:* 1041–1045 (1983).
93 Bellocci, F.; Ansalone, G.; Santarelli, P.; Loperfido, F.; Scabbia, E.; Zecchi, P.; Manzoli, U.: Oral nifedipine in the long-term management of severe chronic heart failure. J. cardiovasc. Pharmacol. *4:* 847–855 (1982).
94 Brooks, N.; Cattell, M.; Pidgeon, J.; Balcon, R.: Unpredictable response to nifedipine in severe cardiac failure. Br. med. J. *281:* 1324 (1980).
95 Elkayam, U.; Weher, L.; McKay, C.; Rahimtoola, S.: Variable, unpredictable and deleterious hemodynamic response to oral nifedipine in severe chronic heart failure: an experience in 32 patients (Abstract). J. Am. Coll. Cardiol. *3:* 478 (1984).
96 Rousseau, M. F.; Pouleur, H.: Calcium antagonism free of negative inotropic effects? A comparison of intracoronary nifedipine and nicardipine (Abstract). Circulation *70:* suppl. II, II–304 (1984).
97 Lahiri, A.; Robinson, C. W.; Tovey, J.; Caruana, M. P.; Kohli, R. S.; Harlow, B. J.; Raftery, E. B.: Intravenous nicardipine in patients with chronic heart failure: a nuclear stethescope study. Post-grad. Med. J. *60:* suppl. 4, pp. 35–38 (1984).
98 Burlew, B. S.; Gheorghiade, M.; Jafri, S. M.; Goldberg, A. D.; Goldstein, S.: Acute and chronic hemodynamic effects of nicardipine hydrochloride in patients with heart failure. Am. Heart J. *114:* 793–804 (1987).

99 Cody, R.J.: Use of invasive ergometric exercise to assess the hemodynamic response to nicardipine. Am. J. Cardiol. *59:* 20J–24J (1987).

100 Ryman, K.S.; Kubo, S.H.; Lystash, J.; Stone, G.; Cody, R.J.: Effect of nicardipine on rest and exercise hemodynamics in chronic congestive heart failure. Am. J. Cardiol. *58:* 583–588 (1986).

101 Lipkin, D.P.; Poole-Wilson, P.A.: Treatment of chronic heart failure – review of recent drug trials. Br. med. J. *291:* 993–996 (1985).

102 Bellinetto, A.; Lessem, J.: Effects of nicardipine hydrochloride in cardiac failure: a dose titration and tolerance study. Curr. ther. Res. *36:* 938–950 (1984).

William J. Mc Kenna, MD, Department of Cardiological Sciences, St. George's Hospital Medical School, Cranmer Terrace, London SW17 ORE (UK)

Van Zwieten PA (ed): Clinical Aspects of Calcium Entry Blockers.
Prog Basic Clin Pharmacol. Basel, Karger, 1989, vol 2, pp 143–154

Treatment of Migraine with Calcium Entry Blockers

Peer Tfelt-Hansen, Lars Edvinsson, Jes Olesen

Department of Neurology, Gentofte Hospital, University of Copenhagen, Hellerup, Denmark; Department of Internal Medicine, University Hospital, Lund, Sweden

Introduction

Despite decades of research, the pathogenesis of migraine is not completely known [1]. Already in 1664 Thomas Willis noted that brain vessels were supplied with nerve fibers. Further research has demonstrated that the human cerebrovascular bed receives three types of perivascular fibers: sympathetic fibers containing noradrenaline and neuropeptide Y; parasympathetic fibers storing acetylcholine, vasoactive intestinal peptide and peptide histidine methionine, and sensory fibers containing inter alia substance P, neurokinin A, calcitonin gene-related peptide [2]. The most prominent symptom in various forms of migraine is headache. Experiments by Ray and Wolff [3] demonstrated in neurosurgical patients that meningeal arteries and large cerebral arteries at the base of the brain were sensitive to mechanical and thermal stimulation which elicited local pain as well as pain referred to various parts of the area supplied by the trigeminal nerve or upper 3 cervical roots. Thus, the trigemino-cerebrovascular pathway is involved in the process of pain. Other types of perivascular nerve fibers may be involved in the case of classic migraine where a localized reduction of cerebral blood flow occurs [4]. The vasoconstrictor response may lead to focal hypoxia/oligemia of such a magnitude that corresponding neurological symptoms are observed [5].

It is no wonder in view of the pathophysiological uncertainty that the mechanism and site of action of antimigraine drugs are poorly understood.

In the following we will first review critically the evidence for prophylactic efficacy in migraine for the calcium entry blockers CEBs flunarizine, nimodipine, verapamil, and diltiazem, and then their possible mode of action and the clinical implications will be discussed.

Flunarizine

Placebo-Controlled Trials

Among the CEBs, flunarizine is the most extensively studied in migraine. In placebo-controlled double-blind trials the results reported so far are unequivocally positive. There are four double-blind trials published concerning the efficacy of flunarizine in adult migraine patients [6–9]. Concerning the prophylactic effect in children an extensive abstract has been published [10]. For an overview of these studies, see table I. Flunarizine was significantly better than placebo in four studies in adults. The effect was mostly or exclusively reflected by a decreased number of attacks, whereas severity of pain and average duration of the attacks were unaltered. The effect of flunarizine was very slow to manifest and increased gradually over 3 months. After 2 months of treatment, the frequency of attacks was approximately reduced to half compared to placebo. In one convincing cross-over study [9], only patients with common migraine were studied; flunarizine, appeared to be effective in that particular subform of migraine as illustrated in figure 1. In another study, classic migraine was examined [8]. No data concerning the aura symptoms were, however, given, so that the nature and details of the patients remain somewhat obscure. Statistical significance was only reported for a headache index, and the study is hardly convincing, because of the small number of patients, a high drop-out rate and use of the relatively weak parallel-group design. In two studies, mixed populations were studied with the parallel-group design [6, 7]. The number of patients with classic migraine was small precluding definite conclusions concerning efficacy in classic migraine. It, therefore, remains uncertain wheter flunarizine acts only in common migraine or also in classic migraine.

In children flunarizine (5 mg) was found to be superior to placebo in migraine prophylaxis [10] and, as in adults, the effect was reflected by a decreased number of attacks.

Comparative Trials

Flunarizine has been compared to a standard drug, pizotifen, in three studies [11–13, see table I]. The design was quite similar in two of the

Table I. Double-blind trials of flunarizine in migraine prophylaxis

References	Patients n	Type of patients: common	Type of patients: classic	Type of design: group	Type of design: cross-over	Duration of periods weeks	Dose mg/day	Compared to	Results attack frequency reduction
6	58	+	+	+		13	10	placebo	57%, $p<0.01$
7	35	+	+	+		12	10	placebo	53%, $p<0.05$
8	20		+?	+		13	10	placebo	?[1]
9	27	+			+	16	10	placebo	38%, $p<0.01$
10[2]	42	+	+	+		12	5	placebo	66%, $p<0.01$
11	75	+	+	+		18	10	pizotifen 2–3 mg	NS
12	18	+	+		+	12	15	pizotifen 1.5 mg	NS
13	35	+	+	+		18	10	pizotifen 2.2 mg	NS
15	25	?	?	+		12	10	nimodipine 120 mg	?[3]
16	69	+	+	+		16	10	propranolol 120 mg	NS
18[2]	29	+	+	+		12	5–10	aspirin 2–5 mg/kg	NS

[1] Significant difference for a headache index.
[2] Children.
[3] No significant difference for a headache index.

studies [11, 13] using parallel groups and 4-month treatment periods. In neither study was any significant difference detected between the two drugs, which were given in doses of flunarizine (10 mg) and pizotifen (2–3 mg). The first study was rather large (75 patients) but the other clearly too small (35 patients) to attain a reasonable power with parallel groups [14]. Neither study included placebo, and neither gave confidence limits or power estimates. In one study [12], the cross-over design was used and 18 out of 27 patients recruited completed the study, where a high dose of flunarizine (15 mg) was compared to the low normal dose of pizotifen (1.5 mg). Both drugs decreased the frequency of attacks compared to a run-in period but the drugs were not significantly different. Although the evidence thus indicates that there is no major difference between flunarizine and pizotifen, a more definite statement must await larger studies.

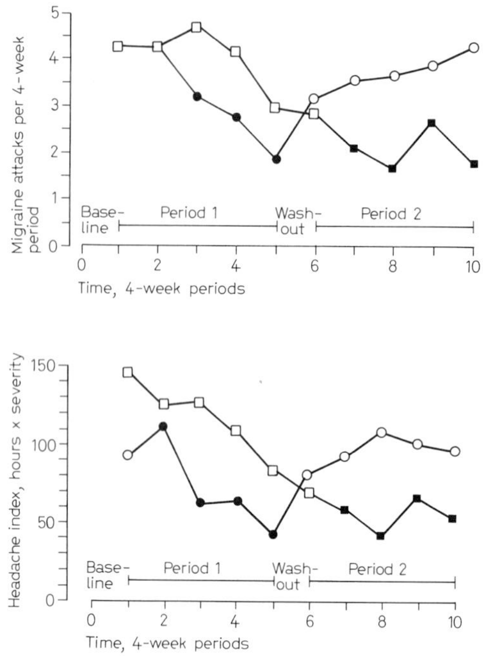

Fig. 1. Evolution in number of migraine attacks per 4 weeks and a headache index during a double-blind trial of flunarizine versus placebo [9]. Mean values are shown. A clear effect of flunarizine is noted with regard to both parameters. ●—○ = Sequence flunarizine – placebo; □—■ = Sequence placebo – flunarizine; ●, ■ = flunarizine treatment.

Flunarizine has also been compared to nimodipine using a parallel-group design involving 25 patients [15]. Both drugs decreased the headache index compared to before treatment. No confidence limits were given and the study was clearly too small to demonstrate comparability [14].

Flunarizine (10 mg) was compared with propranolol (120 mg) in 69 patients with both common and classic migraine in a double-blind study with a parallel-group design [16, see table I]. Both drugs, given for 4 months, decreased the frequency of migraine attacks, and no significant difference was found between the drugs. The study is, as mentioned above, probably too small to demonstrate comparability, but it is further partly invalidated by the use of propranolol (120 mg) instead of the normal propranolol (160 mg) used in most prophylactic trials with [17]. Further studies comparing flunarizine and a β-blocker are therefore needed.

In a double-blind study with a parallel-group design, flunarizine (5–10 mg) and aspirin (2–5 mg/kg), given for 3 months, were compared in 29 children with both common and classic migraine [18]. Both drugs decreased attack frequency by more than 50% compared to a run-in period in approxixmately 70% of cases. No confidence limits were given and the study was as mentioned above too small to demonstrate comparability [14]. Furthermore, no placebo control was included.

Other Studies

Two very large nonblinded studies have evaluated long-term results and side effects of flunarizine [19, 20]. The drug was generally well tolerated, and the favorable results after 3 months were well maintained and in fact further improved after 1 and 2 years. Major side effects were drowsiness, weight gain and depression. Some of these side effects may not have been caused by flunarizine, since double-blind trials have only shown minor differences between flunarizine and placebo with regard to side effects.

Nimodipine

Placebo-Controlled Trials

Two studies deal with a comparison between nimodipine and placebo in migraine prophylaxis [21, 22]. They both used the same dose of nimodipine (120 mg) and both were positive (table II). The effect seems to be predominantly reflected by a decreased number of attacks. The first study comprised mostly patients with common migraine (38/50) and the second mostly those suffering from classic migraine (20/33). The materials are too small to allow separate calculations of the results in each subform. However, an interesting finding in the second study [22] was that the duration of the aura phase with neurological symptoms in classic migraine was of significantly shorter duration during treatment with nimodipine (mean duration 19 min) than during placebo treatment (mean duration 49 min).

Comparative Trials

Recently, nimodipine (120 mg) was compared with pizotifen (1.5 mg) in a cross-over study in 43 patients [23]. Both drugs decreased the number of attacks compared with a placebo run-in period, and there were fewer side

Table II. Double-blind trials of nimodipine in migraine prophylaxis

References	Patients	Type of patients		Type of design		Duration of periods	Dose	Compared to	Results attack frequency reduction
	n	common	classic	group	cross-over	weeks	mg/day		
21	50	+	+	+		13	120	placebo	p<0.05
22	29	+	+		+	8	120	placebo	28%, p<0.01
23	43	+	+		+	12	120	pizotifen 1.5 mg	NS
25	20	+			+	13	120	pizotifen 1.5 mg	?

effects during the nimodipine period than during the pizotifen period. The authors claimed that the trial demonstrated that both drugs were significantly better than placebo. The comparison, however, of the treatment periods with a placebo run-in period is not strictly correct since a so-called 'time effect', that is a decrease in frequency irrespective of treatment with time [24], might be responsible for the observed effect. In an extended abstract with a similar design in 20 patients nimodipine and pizotifen were found equally effective [25; table II]. As mentioned above, nimodipine was found equally effective as flunarizine in a small pilot study [15; table I]. These comparative trials include only a limited number of patients and no confidence limits are given. Statements regarding the comparability of nimodipine to other prophylactic drugs in migraine must await further studies.

Other Studies

Nimodipine (60 mg daily) was compared to nimodipine, 120 mg daily [26]. Methodological information was scarce. Compared to baseline, both regimens reduced the frequency of attacks. Since patients often improve with time regardless of treatment [24], such a comparison is of limited interest.

Verapamil

In two cross-over, double-blind, placebo-controlled studies [27, 28] verapamil (320 and 240 mg daily) seemed to have a prophylactic effect in migraine. Both studies are, however, invalidated by unacceptably high drop-out rates, 11 out of 23 patients recruited [27], and 16 out of 28 [28]. These two trials can thus not constitute satisfactory evidence in favor of a migraine prophylactic effect of verapamil.

Diltiazem

Diltiazem (90 mg daily) given to 15 patients with frequent attacks of migraine (>2 per week) decreased the number of attacks by 86% and the severity by 55% [29]. In a pilot study, patients not responding to nadolol had a significant beneficial effect by diltiazem [30]. These observations need confirmation in controlled clinical trials.

Discussion

Review of Trials

The main result of the present critical review of controlled clinical trials of CEBs in migraine prophylaxis is that flunarizine has demonstrated convincing superiority over placebo. The two studies published so far indicate that nimodipine is superior to placebo, but more studies are needed before a final conclusion can be drawn. For verapamil and diltiazem the evidence is insufficient, and these two CEBs will therefore not be discussed further.

Whereas there is good evidence for the prophylactic efficacy of flunarizine per se, the question of comparability with currently used standard treatment with β-blockers and 5-hydroxytryptamine (5-HT) antagonists remains open. Claims of equipotency should always be regarded with suspicion in a controlled clinical trial lacking substantiation by the presentation of narrow confidence limits [17, 31]. In the comparative trials published so far such evidence has not been presented. In order to obtain narrow confidence limits, a considerable number of patients is needed [24]. For practical reasons, comparability can probably only be demonstrated in cross-over studies [24]. In the comparative cross-over study published so far [12], only 18 patients completed the trial, a number not sufficient for demonstrating comparability. Furthermore, in controlled clinical trials on migraine prophylaxis, it is often observed that the patients improve with time regardless of treatment, the so-called 'time effect' [24]. Therefore, in a comparative trial with two presumably active drugs, a placebo-control should also be included in order to demonstrate convincingly that significant improvement compared to baseline is not just the 'time effect'. To demonstrate equipotency between new drugs as the CEBs and standard treatment with, e.g., β-blockers, large placebo-controlled trials using standard doses of the drugs are needed. This kind of trial has not been performed so far, and the question of comparability of CEBs to other treatments remains open.

Mode of Action

When a completely new type of drug is effective in migraine prophylaxis it may shed light on the mechanism underlying migraine attacks. The pathophysiology of migraine is, however, complex, and one should be cautious in theorizing too much. Thus, not all β-blockers are effective in migraine prophylaxis and, when comparing the pharmacological profils of the effective and ineffective β-blockers, the only thing that characterizes the effective β-blockers is the lack of intrinsic sympathomimetic activity [32]. This negative finding has not been helpful in understanding migraine.

The CEBs are potent vasodilators in many vascular beds. Precontracted human brain vessels are effectively relaxed by CEBs [33, 34]. Some studies have purportedly shown that cerebral vessels are more sensitive to CEBs than are peripheral arteries [35]. One reason for this may be that they are highly dependent upon extracellular calcium for their activation. The effectiveness of some CEBs in migraine could therefore be interpreted as evidence for a considerable role of the vasculature in the pathophysiology of migraine.

Flunarizine was introduced for migraine prophylaxis on the theoretical basis of being a protecting agent against hypoxia, which was thought to be present during migraine attacks [36]. While hypoxia may occur during attacks of classic migraine, this is not the case in common migraine where flunarizine has been most convincingly demonstrated to be effective [9]. The cerebral blood flow changes observed during attacks of classic migraine in patients resemble in many ways the blood flow changes seen after the experimental phenomenon spreading depression of Leao in rats [37]. Therefore, the effect of flunarizine on spreading depression in rats has been studied. The results are equivocal, with one study using a high dose of flunarizine (40 mg/kg intraperitonally) suggesting that flunarizine may increase the threshold for spreading depression [38]. Another study, using

Table III. Inhibition by flunarizine and nimodipine of potassium-induced contractions in human cerebral arteries from Jansen et al. [45]

CEB	n	I_{max}	IC_{50}
Flunarizine	4	89 ± 6	$4.6 \pm 3.7 \times 10^{-8} M$
Nimodipine	6	90 ± 4	$2.8 \pm 1.8 \times 10^{-10} M$

Values given are means ± SEM; n = number of vessel segments tested.

oral doses of flunarizine (20 mg/kg) failed to demonstrate any effect on spreading depression [39]. Flunarizine (10 mg) daily results in plasma concentrations of 10–100 ng/ml [40]. Taking the high protein binding (>90%) of flunarizine in plasma into account, the free plasma concentration of flunarizine will be about 10^{-8} to 10^{-9} *M*. As shown in table III, flunarizine inhibits contractions of human cerebral arteries in concentrations in the 10^{-7} *M* range and is therefore unlikely to exert any calcium-antagonistic effect on cerebral arteries in man. A so-called antispastic effect of flunarizine can therefore be ruled out as the mechanism of action. It should be kept in mind that flunarizine, apart from being a CEBs also has an antihistaminic effect [40], and probably also has some effect on neurotransmittor systems in man as indicated by its antiepileptic efficacy [41]. From this brief discussion of possible modes of action of flunarizine it is probably clear that knowledge concerning the total pharmacological profile of the drug and concerning migraine pathophysiology is insufficient to make a qualified guess about the mode of action of flunarizine in migraine.

Nimodipine was introduced as a prophylactic agent in migraine, because it has a selective action on cerebral vessels and, as shown in table III, nimodipine, which is present in plasma in concentrations in the 10^{-8} *M* range [42], is likely to exert a calcium-antagonistic effect on cerebral arteries in man. There is, however, also experimental evidence that nimodipine can affect neurotransmittors [43]. Several modes of action are thus possible for nimodipine, but its implications for migraine pathophysiology will, however, have to wait until the efficacy of nimodipine is confirmed.

Clinical Implications

Despite the reservations made above concerning the comparability of the CEBs to standard prophylactic treatment in migraine, there is no doubt that the addition of flunarizine and perhaps nimodipine to our armentarium for the treatment of migraine is a major advantage.

Firstly, patients who do not tolerate previous treatments (e.g., asthmatic patients with migraine who cannot tolerate β-blockers) can now be given the CEB flunarizine. Secondly, patients not responding to standard treatment might respond to this new type of drugs, and in the future we might be able to analyze the responses in large double-blind studies and predict which type of drug, amine antagonists,β-blockers, or CAs, is most likely to be beneficial in special subgroups of migraine patients. Even if the execution of such large-scale responder-finding studies is cumbersome and costly, the identification of responder subgroups will be of so considerable benefit to the patients that costs and efforts will be easily justified. Thirdly, it is still possible that CEBs may be proven more effective than previous treatments and/or that they have fewer side effects. Finally, a combination with β-blockers may prove more effective than either drug alone, as has been shown in stable angina pectoris [44].

References

1 Olesen, J.; Edvinsson, L.: Basic mechanisms of headache, pp. 1–509 (Elsevier, Amsterdam 1988).

2 Edvinsson, L.; MacKenzie, E.T.; McCulloch, J.; Uddman, R.: Perivascular innervation and receptor mechanisms in cerebrovascular bed; in Wood, Cerebral blood flow, pp. 145–172 (McGraw-Hill, New York 1987).

3 Ray, B.S.; Wolff, H.G.: Experimental studies on headache: pain-sensitive structures of the head and their significance in headache. Archs Surg. *41:* 813–856 (1940).

4 Olesen, J.; Larsen, B.; Lauritzen, M.: Focal hyperemia followed by spreading oligemia and impaired activation of rCBF classic migraine. Ann. Neurol. *9:* 344–352 (1981).

5 Skyhøj Olsen, T.; Olesen, J.: Regional cerebral blood flow in migraine and cluster headache; in Olesen, Edvinsson, Basic mechanisms of headache (Elsevier, Amsterdam, 1988).

6 Louis, P.: A double-blind placebo-controlled prophylactic study of flunarizine in migraine. Headache *21:* 235–239 (1981).

7 Frenken, C.W.G.M.; Nuijten, S.T.M.: Flunarizine, a new preventive approach to migraine. Clin. Neurol. Neurosurg. *86:* 17–20 (1984).

8 Mendenopoulos, G.; Manafi, T.; Logothesis, I.; Bostantjopoulou, S.: Flunarizine in the prevention of classical migraine: a placebo-controlled evaluation. Cephalalgia *5:* 31–37 (1985).

9 Sørensen, P.S.; Hansen, K.; Olesen, J.: A placebo-controlled, double-blind, cross-over trial of flunarizine in common migraine. Cephalalgia *6:* 7–14 (1986).

10 Sorge, F.; DeSimone, R.; Marano, E.; Orefice, G.; Carrieri, P.: Efficacy of flunarizine in the prophylaxis of migraine in children. Cephalalgia *5:* suppl. 3, p. 172 (1985).

11 Louis, P.; Spierings, E.L.H.: Comparison of flunarizine and pizotifen in migraine treatment: a double-blind study. Cephalalgia *2:* 197–203 (1982).
12 Cerbo, R.; Casacchia, M.; Formisano, R.; Feliciani, M.; Cusimano, G.; Buzzi, M.G.; Agnoli, A.: Flunarizine-pizotifen single-dose double-blind cross-over trial in migraine prophylaxis. Cephalalgia *6:* 15–18 (1986).
13 Rascol, A.; Montastruc, J.-L.; Rascol, O.: Flunarizine versus pizotifen: a double-blind study in the prophylaxis of migraine. Headache *26:* 83–85 (1986).
14 Tfelt-Hansen, P.; Nielsen, S.L.: Patient numbers needed in prophylactic migraine trials. Neuroepidemiology *6:* 214–219 (1987).
15 Bussone, G.; Baldini, S.; D'Andrea, G.; Cananzi, A.; Frediani, F.; Caresia, L.; Ferro Milone, F.; Boiardi, A.: Nimodipine versus flunarizine in common migraine: a controlled pilot study. Headache *27:* 76–79 (1987).
16 Soyka, D.; Oestreich, W.: Flunarizine versus Propranolol in der Intervall-Prophylaxe der Migräne – Eine multizentrische Doppelblindstudie in 12 Kliniken. Nervenheilkunde *6:* 177–183 (1987).
17 Tfelt-Hansen, P.: Efficacy of β-blockers in migraine. A critical review. Cephalalgia *6:* suppl. 5, pp. 15–24 (1986).
18 Pothmann, R.: Migräneprophylaxe mit Flunarizin und Acetylsalizylsäure. Eine Doppelblindstudie. Mschr. Kinderheilk. *135:* 646–649 (1987).
19 Bono, G.; Manzoni, G.C.; Martucci, N.; Baldrati, A.; Farina, S.; Cassabgi, F.; DeCarolis, P.; Nappi, G.: Flunarizine in common migraine. Italian cooperative trial. II. Long-term follow-up. Cephalalgia *5:* suppl. 2, pp. 155–158 (1985).
20 Leandri, M.; Parodi, C.; Bacigalupo, F.; Farinini, D.: Long-term study on effectiveness of flunarizine in headache prophylaxis. Cephalalgia *5:* suppl. 3, pp. 556–557 (1985).
21 Gelmers, H.J.: Nimodipine, a new calcium antagonist, in the prophylactic treatment of migraine. Headache *23:* 106–109 (1983).
22 Havanka-Kannianinen, H.; Hokkanen, E.; Myllylä, V.V.: Efficacy of nimodipine in the prophylaxis of migraine. Cephalalgia *5:* 39–43 (1985).
23 Havanka-Kannianinen, H.; Hokkanen, E., Myllylä, V.V.: Efficacy of nimodipine in comparison with pizotifen in the prophylaxis of migraine. Cephalalgia *7:* 7–13 (1987).
24 Olesen, J.; Krabbe, A.E.; Tfelt-Hansen, P.: Methodological aspects of prophylactic drug trials in migraine. Cephalalgia *1:* 127–141 (1981).
25 Micieli, G.; Trucco, M.; Agostinis, C.; Mancuso, A.; Papalia, F.; Sinforiani, E.: Nimodipine vs. pizotifen in common migraine: results of a double-blind cross-over trial. Cephalalgia *5:* suppl. 3, pp. 532–533 (1985).
26 Meyer, J.S.; Hardenberg, J.: Clinical effectiveness of calcium entry blockers in prophylactic treatment of migraine and cluster headache. Headache *23:* 266–277 (1983).
27 Solomon, G.D.; Steel, J.G.; Spaccavento, L.J.: Verapamil prophylaxis of migraine. J. Am. med. Ass. *250:* 2500–2502 (1983).
28 Markley, H.G.; Cheronis, C.D.; Piepho, R.W.: Verapamil in prophylactic therapy of migraine. Neurology *34:* 973–976 (1984).
29 Riopelle, R.J.; McCans, J.L.: A pilot study of the calcium antagonist diltiazem in migraine syndrome prophylaxis. J. can. Sci. Neurol. *9:* 269 (1982).
30 Smith, R.; Schwartz, A.: Diltiazem prophylaxis in refractory migraine. New. Engl. J. Med. *310:* 1327–1328 (1984).
31 Tfelt-Hansen, P.; Olesen, J.: Methodological aspects of drug trials in migraine. Neuroepidemiology *4:* 204–226 (1985).

32 Weerasuriya, K.; Patel, L.; Turner, P.: β-adrenoceptor blockade and migraine. Cephalalgia *2:* 33–45 (1982).

33 Edvinsson, L.; Brandt, L.; Andersson, K.-E.; Bengtsson, B.: Effect of a calcium antagonist on experimental constriction of human brain vessels. Possible efficacy in cerebrovascular spasm. Surg. Neurol. *11:* 327–330 (1979).

34 Brandt, L.; Andersson, K.-E.; Edvinsson, L.; Ljunggren, B.: Effects of extracellular calcium and of calcium antagonists on the contractile responses of isolated human pial and mesenteric arteries. J. cereb. Blood Flow Metab. *1:* 339–347 (1981).

35 Peroutka, S.J.: Relative potency and selectivity of calcium antagonists used in the treatments of migraine. Headache *25:* 55–58 (1984).

36 Amery, W.K.: Brain Hypoxia: the turning-point in the genesis of the migraine attack? Cephalalgia *2:* 83–109 (1982).

37 Lauritzen, M.: Cortical spreading depression as a putative migraine mechanism. T.I.N.S. *10:* 8–13 (1987).

38 Marrannes, R.; Wauquier, A.: Influence of flunarizine on spreading depression. Cephalalgia *5:* suppl. 3, p. 208 (1985).

39 Hansen, A.J.; Lauritzen, M.; Tfelt-Hansen, P.: Spreading cortical depression and antimigraine drugs; in Amery, van Nueten, J.M.; Wauquier, The pharmacological basis of migraine therapy, pp. 161–170 (Pitman, London 1984).

40 Holmes, B.; Brogden, R.N.; Heel, R.C.; Speight, T.M.; Avery, G.S.: Flunarizine. A review of its pharmacodynamic and pharmacokinetic properties and therapeutic use. Drugs *27:* 6–44 (1984).

41 Overweg, J.; Binnie, C.D.; Meijer, J.W.A.; Meinardi, H.; Nuijten, S.T.M.; Schmalz, S.; Wauquier, A.: Double-blind placebo-controlled trial of flunarizine as add-on therapy in epilepsy. Epilepsia *25:* 217–222 (1984).

42 Allen, G.S.; Ahn, H.S.; Preziosi, T.J.; et al.: Cerebral arterial spasm – controlled trial of nimodipine in patients with subarachnoidal haemorrhage. New. Engl. J. Med. *308:* 619–624 (1983).

43 Itil, T.M.; Itil, K.Z.: The comparative CNS pharmacology of nimodipine in humans; in Betz, Deck, Hoffmeister Nimodipine. Pharmacological and clinical properties, pp. 185–202 (Schattauer, Stuttgart 1985).

44 Leon, M.B.; Rosing, D.R.; Bonnow, R.O.; Epstein, S.W.: Combination therapy with calcium-channel blockers and β-blockers for chronic stable angina pectoris. Am. J. Cardiol. *55:* 69–80 (1985).

45 Jansen, I.; Tfelt-Hansen, P.; Edvinsson, L.: Comparison of the vasomotor effects of the calcium entry blockers flunarizine and nimodipine in human cerebral and temporal arteries. Cephalalgia *5:* suppl. 3, pp. 206–207 (1985).

Peer Tfelt-Hansen, MD, Department of Neurology, Gentofte Hospital,
University of Copenhagen, DK–2900 Hellerup (Denmark)

Van Zwieten PA (ed): Clinical Aspects of Calcium Entry Blockers.
Prog Basic Clin Pharmacol. Basel, Karger, 1989, vol 2, pp 155–193

Treatment of Neurological Disorders Except Migraine with Calcium Antagonists

With 1 Color Plate

A. Hartmann[a], *P.A. van Zwieten*[b]

[a]Universitäts-Nervenklinik, Neurologie, Bonn, BRD; [b]Divisions of Pharmacotherapy and Cardiology, Academic Medical Centre, University of Amsterdam, The Netherlands

Events at the Cellular Level Associated with Cerebral Ischemia

Typical characteristics of neuronal metabolism are a high adenosine triphosphate (ATP) turnover, a lack of oxygen reserve, only limited glucose and glycogen storage and low concentrations of high-energy compounds. Therefore, ATP turnover depends largely on a sufficient supply of oxygen and glucose [1, 2]. The cerebral metabolic rate of oxygen is about 3.5 ml/100 g/min and is the basis for neuronal activity and preservation of cellular integrity. The glucose metabolic rate approximately amounts to 4.5 ml/100 g/min [1]. Glucose storage exceeds that of oxygen, which explains the high sensitivity of the brain to hypoxia, compared to a limited resistance to glucose deficiency [3]. If tissue perfusion is insufficient to guarantee oxygen supply in response to the metabolic demand, aerobic metabolism may be switched to anaerobic processes with increased production of lactic acid. Reduction of cerebral blood flow (CBF) is followed by an increased oxygen extraction, but after reaching the maximal oxygen extraction, ischemia with functional failure takes place [4].

Cerebral ischemia may be divided into incomplete and complete ischemia. The functional outcome of the tissue with incomplete ischemia depends on residual perfusion through the original or through anastomotic vessels, on oxygen supply and on the presence of high-energy phosphates. It has been suggested that a high glycogen level prior to the onset of ischemia may endanger functional recovery due to increased lactate concentration [5], which may be less evident during complete ischemia [6, 7]. However, this view has been confirmed by only few clinical studies [8]. Usually, metabolic disturbances may be observed after blood flow has decreased to a level of 45–50% below the steady-state value [9]. The time period after which irreversible tissue damage is observed may be prolonged from 4 to 6 min to a several-fold value [10]. The most detrimental effect has been attributed to the lactic acid level, which may rise up to 40 μM/g tissue during incomplete ischemia and to only 20 μM/g tissue during complete ischemia [11]. This points to the observation that complete cerebral ischemia may be tolerated better than incomplete ischemia [12, 13]. Experimental neurologic recovery may be observed even after 15–25 min of complete cessation of blood circulation in the brain [14, 15]. Recovery time for EEG, evoked potentials and some metabolic parameters could be prolonged even to 60 min after complete ischemia [16].

2–3 min after complete ischemia, the ATP level is reduced to almost zero, whereas the concentration of adenosine diphosphate increases gradually. Alternative reactions for the preservation of metabolism, like the creatinine-phosphokinase reaction or anaerobic glycolysis of glycogen or glucose with formation of lactic acid compensates for the decrease of ATP only for a short time.

The initial decrease of ATP concentration in complete ischemia is rapid and slows down after 60 s [17, 18]. Finally, ATP approaches zero [19, 20], whereas during incomplete ischemia ATP may be established even after 24 h.

1–2 min after complete cessation of blood circulation to the brain, an efflux of potassium (K^+) may be recorded [21–23]. At the same time an enhanced influx of calcium (Ca^{2+}) occurs [24]. In addition, Ca^{2+} is released from the endoplasmic reticulum. The concomitant influx of sodium (Na^+) and water leads to significant brain swelling. The recovery of brain tissue function after cerebral ischemia depends largely upon the intensity and duration of CBF reduction. The concept of flow thresholds [25] has led to the suggestion that a narrow window might exist which permits early tissue function recovery if blood flow is normalized within a short time. The functional failure does not occur at once, but progresses slowly over a period of a few to several minutes, depending on other parameters like oxygen supply, anastomotic network or, rheologic parameters. Figure 1 indicates the cortical-evoked potentials of a baboon in which the acute and total clipping of the middle cerebral artery led to abolition of these potentials after 16 min. This is partially due to the maturation phenome-

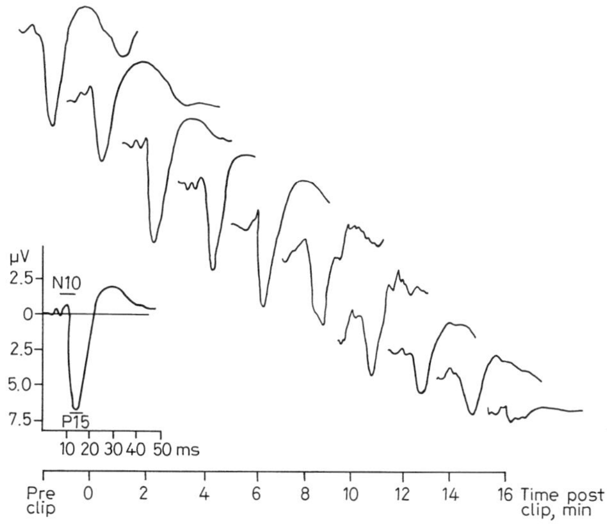

Fig. 1. Cortical-evoked potentials over the postcentral gyrus after stimulation of the median nerve in a baboon before and after clipping the middle cerebral artery. The evoked responses are completely abolished after 16 min.

non. Electrical cortical failure in humans starts at a hemispheric flow level of about 16–18 ml/100 g/min. Below this level, alterations of the EEG will reflect the occurrence of functionally significant ischemia [26]. In monkeys, evoked somatosensory potentials from the cortex decrease at a flow of about 15 ml/100 g/min [25]. With respect to the EEG as an indirect indicator of tissue metabolism and its dependency on CBF, it has been shown that (during halothane-nitrous oxide-oxygen anesthesia) the EEG is altered irreversibly below a flow level of 10 ml/100 g/min [27]. Massive efflux of K^+ and influx of Ca^{2+} occurs below a flow of 10 ml/100 g/min [25], which again marks irreversible tissue damage. The zone between the occurrence of functional failure and irreversible damage, as indicated by electrical silence and moderate astrocyte and neuronal swelling, again due to an impairment of oxygen transport within the tissue, is the so-called 'penumbra'. The penumbra represents tissue characterized by a loss of functional capacity but preservation of structural integrity. Functional recovery may be achieved by improvement of CBF.

Regulatory processes such as autoregulation (which guarantee the independence of CBF from the perfusion pressure) and the CO_2 reactivity might fail in the area of the penumbra. This indicates a high sensitivity of the tissue to any changes in blood pressure, intracranial pressure or local tissue pressure as a result of brain edema. Whereas in regional ischemia any definite concept concerning the dependence of functional capacity on the intensity and duration of ischemia remains uncertain, it seems justified, in complete acute ischemia, to estimate the duration of the EEG survival time to 15–20 s, that of the cortical-evoked potentials to 3–4 min and that of the membrane polarization of the neurons to 6–8 min [28]. In regional ischemia, cellular swelling (due to breakdown of membrane K^+–Na^+–AT-Pase) with narrowing of the extracellular space marks the beginning of the cytotoxic edema and occurs at a flow level between 10 and 16 ml/100 g/min. K^+ efflux occurs at lower flow levels (8–10 ml/100 g/min), that is at the same level where the EEG is completely suppressed [29–33]. Initially, the K^+ efflux is rapid and then becomes slower. The K^+ concentration increases from 3 to 20 mEq/l within 5 min in the extracellular space and at the same time Na^+ decreases from 140 to 110 mEq/l. After 60 min of ischemia, the extracellular K^+ concentration amounts to about 56 and that of the Na^+ to about 55 mEq/l [30]. The intracellular swelling (first observed in the gray matter) leads to a narrowing of the extracellular space by approximately 50%. The total water volume does not change during this period of cellular edema, the blood-brain barrier is still undamaged.

The second step development of brain edema occurs at a later stage, when (during recirculation or during prolonged partial ischemia) the blood-brain-barrier breaks down. Macromolecules from the blood stream enter the perivascular space, which again increases the extracellular water content [34]. The first events which can be detected morphologically with respect to the development of damage to the blood-brain-barrier occur after 3–20 h, depending on the duration of the initial ischemia [35–37].

With the start of recirculation, K^+ is transported into the blood, Na^+ accumulates in the extracellular space and this is accompanied by a massive water accumulation, leading subsequently to an increase in intracranial pressure [38]. Simultaneously, the cerebral lactacidosis leads to vasodilation with brain swelling, which again potentiates the increase in intracranial pressure. This is one of the detrimental factors for the development of the no-reflow phenomenon [39].

The phenomena described above suggest that both recirculation after ischemia and incomplete ischemia may contribute to an unfavorable clinical development. Since this concept casts a negative reflexion on the therapeutic concept of early CBF improvement in cerebral ischemia, other hypotheses should be discussed.

Apart from the aforementioned processes and parameters, loss of calcium homeostasis is also assumed to contribute to ischemic neuronal necrosis. As in nonneuronal tissues, the extra/intracellular gradient of calcium ions is approximately 10,000-fold. Accordingly, a considerable electrochemical force tends to drive calcium ions into the cell. However, adequate regulatory mechanisms via specific membrane calcium channels ensure the maintenance of the extra/intracellular gradient. Apart from the calcium ion concentration in the cytosol, which is functionally of utmost importance but very low ($10^{-7}M$), several subcellular particles, like the sarcoplasmic reticulum, the inner side of the sarcolemma, mitochondria and microsomes, possess the ability to bind and release calcium ions and are involved in the regulation of calcium homeostasis.

Ischemia is associated with a marked decrease in extracellular calcium ions, which is assumed to be triggered by the sharp rise in extracellular potassium level and corresponding depolarization of the cell membrane. Simultaneously, neurotransmitter release like that of noradrenaline into the extracellular space also occurs. Both the depolarization and the transmitter release will trigger the opening of potential-operated (POC) and receptor-operated channels (ROC), respectively. The ROC are assumed to be triggered by α-adrenoceptors (in particular of the α_2-subtype) and hence activated by endogenous noradrenaline and adrenaline. Most of the calcium which rapidly disappears from the extracellular space during ischemia passsively flows into the cells and, as a consequence, the intracellular calcium concentration rises to values up to approximately 10^{-5} *M*. Apart from the influx of calcium ions through regular, specific channels (POC and ROC), a nonspecific influx of Ca^{2+} ions, not necessarily following specific channels, is also induced by severe ischemia. The excessive calcium concentration accumulating in the cytosol as a result of ischemia is believed to accelerate the noxious influences of ischemia by impairing the extracellular ATP build-up via synthesis and also causing inactivation of ATP by complex formation with calcium ions and precipitation. Accordingly, it is generally assumed that the excessive cellular calcium concentration plays a major part in the development of necrosis during ischemia. With respect to ischemic brain damage, however, the evidence for this generally assumed mechanism so far remains circumstantial and a definite proof is still lacking. The therapeutic efficacy of calcium antagonists (CAs) both in experimental models and in conditions of ischemia associated with neurological disorders strongly supports the pathophysiological relevance of high intracellular calcium concentrations for ischemic brain damage.

Subarachnoid Hemorrhage

Pathophysiology and Biochemistry

Patients who survive the initial stage of subarachnoid hemorrhage (SAH) are subject to a substantial degree of mortality and morbidity. Besides acute extraventricular cerebrospinal fluid pathway obstruction, rebleeding, intracranial bleeding and development of brain edema due to different causes, delayed cerebral vasospasms are the most important complications of SAH. Major causes of death in the first week after the initial bleeding are recurrent hemorrhage and vasospasms [40, 41]. Spasms have been observed in about 33–50% of the patients with SAH [42]. Vasospasm is determined by a local or generalized active vasoconstriction and is proven by means of angiographic techniques. Transcranial Doppler sonography is able to indicate alterations of blood flow velocity of the major intracranial arteries following local vasoconstriction and is therefore a suitable procedure for follow-up studies when the existence of vasospasm has been proven.

The initial bleeding of SAH leads to a reduction in CBF in clinical as well as in experimental studies [43–46]. A secondary drop in CBF may be caused by the complicating factors of SAH. In the state of acute SAH autoregulation might be impaired and, therefore, tissue perfusion directly depends on the local perfusion pressure. Vasoconstriction due to vasospasms leads to a local increase in vascular resistance and this is compensated for only when autoregulation in the postspastic territory initiates vascular dilatation. Focal neurologic deficits develop when CBF drops below the already described flow threshold. However, the presence of autoregulation of anastomotic vessels may prevent the decrease in local CBF despite the presence of vasospasm, and partial preservation of autoregulation has been shown in experimental SAH in baboons [47]. This is the cause of the high correlation between vasospasms and neurologic deficits in many, but not all, patients (fig. 2). Vasospasms significantly contribute to the clinical condition of patients with SAH. Vasospasms of the initial phase, shortly after SAH, are unimportant for the clinical outcome, whereas vasospasms of the second phase may be clinically

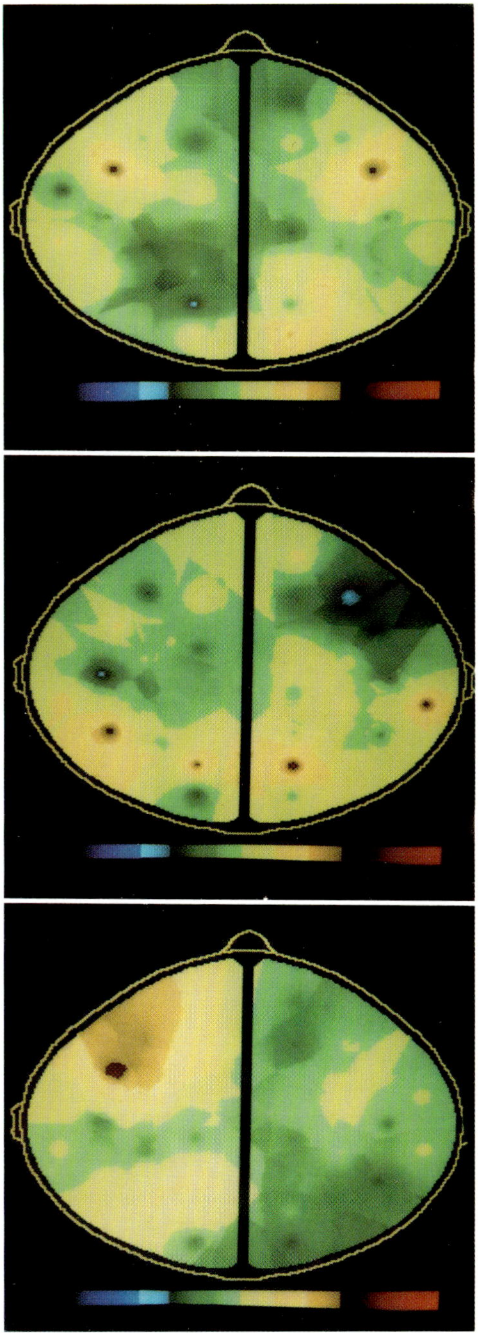

a

b

c

Fig. 2. For legend see reverse side.

Fig. 2. CBF mapping in a patient with right-sided aneurysm of the middle cerebral artery and localized vasospasms. Yellow indicates normal flow and green reduced flow compared to hemispheric mean. *a* CBF on day of SAH. *b* CBF map on day 3 with significant reduction of CBF over the proximal area of the middle cerebral artery (arrow). *c* Spread of ischemia (green) to most of the cortex of the right hemisphere due to extensive vasospasms on day 7.

relevant [42]. The clinical manifestation usually appears between the 5th and the 7th day after the initial bleeding. Vasospasms have been observed only rarely before the 3rd and after the 14th day.

To minimize the risk of recurrent bleeding and development of vasospasms, many neurological centers attempt to achieve an early operation to clip the vascular malformation and even to remove the blood from the subarachnoid space [48–53]. However, even after early surgery, vasospasms with significant clinical influence have been noted [50–53]. Furthermore, many patients in whom an aneurysm has been detected cannot be subjected to surgery.

A close correlation between the amount of blood in the subarachnoid space and the intensity of spasms has been described [54, 55], and early removal of blood may decrease the chance of development, or at least the intensity, of vasospasms [56, 57]. This, however, holds true only for patients in good clinical condition at the time of surgery [58, 59]. The best results are obtained when surgery is performed on the 1st or 2nd day after the first bleeding [60, 61]. Since early surgery is not always possible in SAH and since even surgical intervention as such cannot always prevent vasospasms, other therapeutic measures have been attempted in order to neutralize or avoid vasospasms and their clinical sequelae. Most medical attempts have aimed at neutralizing either the vasoconstriction or at abolishing the primary biochemical cause of vasospasms [62]. Despite the many biochemical factors which might be involved in triggering vasospasms [63–68], the exact pathophysiologic chain remains obscure [69]. The most relevant factors are probably degradation products of the blood or even the red cells [70].

The early state after SAH is characterized by activation of blood coagulation with subsequent release of thromboxane A_2, serotonin and prostaglandin A_2. Thromboxane A_2 and serotonin are potent vasoconstrictors which will be spread by the cerebrospinal fluid. Both factors may contribute to the early phase of SAH. However, vasospasms can be provoked by different substances like whole blood [71, 72], hemoglobin [73–76], histamine [77], potassium chloride [78], norepinephrine [79–83], prostaglandin-2-alpha [78, 84] or serotonin [79, 85, 86]. The late phase of vasospasms might depend on degradation products of erythrocytes like oxyhemoglobin [62–64] and free radicals which activate vasoconstricting prostaglandins [65]. Thrombin may facilitate the recurrence of early and delayed spasms by direct vasoconstriction and activation of thromboxane A_2 and serotonin [66, 67]. The hypersensitivity to vasoconstricting effects of catecholamines has been described for day 3–8 following the episode of SAH [68].

There seems to be a correlation between the occurrence of vasospasms and the amount of blood which has entered the subarachnoid space [87–89].

Factors released by the blood clot, the surrounding tissue or the autonomic nervous system probably contribute to vasoconstriction. This initial effect is followed by an increase in cerebrovascular muscle tone, and depends highly on the entry of calcium through specific channels into the vascular smooth muscle [90, 91]. In addition to the initial biochemical vasoconstriction, which might initiate ischemia, morphological changes occur such as the loss of endothelium, as described in myocardial infarction [92]. Erythrocytes [93] and eventually fibrinogen degradation products or platelet-derived substances may play an important role in the morphological changes occurring in the vascular wall after exposure to blood [94]. Increase in platelet binding to the damaged area intensifies the vasoconstriction [95]; free radicals may play a further important role in this process [96, 97]. The resulting inflammatory reaction in the vascular wall can potentially be neutralized by anti-inflammatory agents [98, 99]. Even if factors inducing chronic spasms of blood-derived factors have not been clearly identified, free Ca^{2+} is always required for the vascular constriction, and the intensity of vasoconstriction directly depends on the concentration of free Ca^{2+} at the ATP in the actin and myosin filaments [100]. The Ca^{2+} responsible for intracranial arterial constriction originates from the extracellular rather than intracellular pool [101–104]. A steady constriction – which reflects the vasospasm – needs a constant replenishment of the intracellular Ca^{2+} pool by influx from the extracellular compartment. The blockade of the increased Ca^{2+} uptake neutralizes the constriction of isolated pial arteries in various species [105–107].

Prevention of Vasospasms

It therefore seems logical to try to prevent vasospasms in patients and to neutralize the clinical effects of existing vasospasms. Several medical approaches have been attempted, all of them without persisting effects (table I). With respect to the therapeutic regimens, it seems appropriate to antagonize even the initial vasoconstriction, which depends more strongly on the calcium influx than the second phase [130, 131]. In monkeys, pretreatment with diltiazem in experimental SAH is more effective than treatment more than 12 h after exposure of the subarachnoid space to

Table I. Therapeutic regimens without any significant preventive effects on vasospasms following SAH

Stimulation of β-adrenergic receptors (isoproterenol)
Nylidrin
Blockade of α-adrenergic receptors (phenoxybenzamine)
Phentolamine
Reserpine
Atropine vagolysis
Phosphodiesterase inhibition (theophylline)
Bradykinin
Dimethylsulfoxide
Inactivation of red cells by intrathecal haptoglobin
Direct vasodilatation by naftidrofuryl, papaverine, prostacyclin, sodiumnitroprusside
Barbiturates
Hemodilution with dextran or mannitol

autologous blood [132]. The same effects were observed with other substances such as aminophylline, papaverine and nifedipine [133–135] although beneficial effects with aminophylline have been shown in some experiments [136] and clinical studies [137–139]. The acute vasoconstriction in experimental protocols is counteracted by CAs like diltiazem [140–143], nicardipine [144, 145], nimodipine [146], nifedipine [147, 148] and verapamil [149]. In addition, flunarizine and lidoflazine both impair Ca^{2+} accumulation in the myofilaments of the cerebral vascular muscle cell [150].

Since nimodipine is claimed to relax intracranial artery tone more strongly than peripheral vessel tone [108–111], particular interest was focussed on this compound [105, 112]. Nimodipine is a lipid-soluble 2,4-dihydropyridine, which, in contrast to nifedipine, readily passes the blood-brain barrier [113, 114]. It impairs the influx of extracellular Ca^{2+} into vascular smooth muscle cells, particularly those of the brain [115, 116]. The constriction following exposure of cerebral vessels to thromboxane, thrombin, KCl, serotonin and phenylephrine is impaired by nimodipine [116–119]. In experimental SAH, it reduces vasospasms [120–122]. In primates, nimodipine (2 μg/kg body weight/min) increases CBF by about 20%; this increase is accompanied by a reduction in blood pressure of approximately 11% [123]. Sahlen et al. [124] have described an increase of CBF of 25–30% following nimodipine administration at doses of about 0.1 μg/kg/min in both normal baboons and in baboons with SAH and this increase of CBF by nimodipine has been confirmed by others [125–127]. Svendgaard et al. [128] elevated blood pressure and observed that nimodipine increases CBF, thus impairing autoregulation and physiological vasoconstriction. This was also shown in cats by Gaab et al. [129].

A breakthrough was achieved by a multicenter study in patients with SAH. The patients were assigned randomly either placebo or oral nimodipine (0.35 mg/kg every 4 h for 21 days) therapy [151]. In all 125 patients, SAH was confirmed by cranial computerized tomography or CSF sampling and only patients with proven aneurysms (as shown by angiography) were included in the study. At the end of the protocol, 8 patients of the placebo group had either died or were severely incapacitated, whereas only 1 patient in the nimodipine group had severe neurologic deficits. Patients with good recovery in the nimodipine protocol showed a significantly higher blood level in the CSF (as judged from cranial computerized tomography) than control patients with good recovery. 12 patients (13%) died during the study, which is significantly less than in patients with vasospasms and no therapy [48, 152–154]. In 62.1% of the grade IV patients and 50% of the grade V patients the outcome was judged 'favorable'. The clinical response was better in patients in whom therapy was started within 24 h after the onset of hemorrhage.

It might be concluded that the outcome of patients in this protocol was better than in those not subjected to therapy. However, both blood pressure reduction and cardiac arrhythmias have been observed as well. Both are factors which may contribute to the spontaneous course and to the course during therapy of existing vasospasms in SAH [55–157]. Transient alterations of transaminase levels (GOT and GPT) are probably due to the solvent, which contains ethanol.

The question arises whether prophylactic nimodipine therapy in patients with SAH is of benefit to *all patients,* irrespective of their clinical grading. The potential dilation of vessels with normal and with pathologic vasomotor control mechanisms, which may lead to an increase of intracranial pressure [128], underlines the necessity of a careful selection of patients. In patients in whom vasomotor control is impaired, a steal phenomenon and impairment of autoregulation cannot be ruled out [129]. However, the theoretical considerations are so far refuted by the clinical experience: Seiler et al. [158] have observed a lesser beneficial effect of nimodipine in patients with SAH and low surgical risk ('risk' as judged from clinical grading and appearance of blood in the cranial computerized tomography), than in those with high risk. The increase in blood flow velocity, as estimated with transcranial Doppler sonography, was increased following the administration of nimodipine [159, 160]. However, it should be pointed out that blood flow velocity does not correlate directly with tissue perfusion, and microcirculatory blood flow might be quite different from intracranial proximal blood flow velocity. Since prevention of clinical symptoms due to cerebral ischemia by nimodipine is more effective than prevention of vasospasms, it was believed that the substance exerts effects other than vasodilatation, like the opening of collaterals [161, 162], action on metabolism [163] and impairment of platelet aggregation, that are believed to cause arterial thrombosis. This impairment of platelet function was shown at least for nifedipine, which, however, is not used as a cerebral antispastic substance [164]. For practical purposes, it may be recommended to administer nimodipine (2 mg/h i.v. over 6–8 days), followed by 6 × 45 mg/day for another week.

The plasma concentrations of about 27 ng/ml after intravenous administration and more than ± 13 ng/ml after oral administration are sufficient to decrease the rate of ischemic complications significantly [165].

Blood pressure should be monitored carefully to avoid any hypoperfusion due to the decrease in systemic blood pressure. Since substances like hydroxyethyl starch do not signyficantly affect blood coagulation [own results, unpubl.] they may be used to maintain perfusion pressure at a constant level, since nimodipine, a vasodilatator, decreases blood pressure in individual cases. Accordingly, slight hypervolemia is induced by 500 ml 6% hydroxyethyl starch over 5 h/day. However, awaiting the outcome of controlled studies, this protocol should be followed only in patients with SAH and clipped aneurysms. In patients with existing vasospasms and clipped aneurysms, this hypervolemic hemodilution may also be followed to compensate for existing ischemia.

Treatment of Existing Vasospasms

Only a few recommendations can be found in the literature with regard to the therapy of neurologic deficits due to already existing vasospasms. Böker et al. [166] have described a reduction in the intensity of vasospasms after infusion of nimodipine into the internal carotid artery in 2 cases and clinical improvement in a 3rd case without control angiography. However, in addition they point out, that in 1 of these patients, the site of the aneurysm, as indicated by contrast medium filling, was increased. They raise the question of the danger incurred by patients undergoing preoperative therapy of vasospasms with vasodilating agents, including the CAs.

Koos et al. [167] have investigated the effect of nimodipine in patients with neurologic deficits due to cerebral vasospasms after SAH in a noncontrolled multicenter study. 91 patients completed the protocol. Clinical signs of cerebral ischemia occurred in all of them, in 60 patients prior to surgery and in 31 postoperatively. At the beginning of the treatment 35 patients were classified as grade IV according to Hunt and Hess's classification and altogether 76 patients (84%) were classified as stages III–V. Patients were treated with 24–48 mg nimodipine/day i. v. for 7–10 days followed by oral doses of up to 240 mg/day for an undefined period. In two thirds of the patients, therapy was started within 24 h after the onset of ischemic symptoms. Angiographic evidence of existing vasospasms was not obtained in all cases. 65% of the patients exhibited complete recovery of the ischemic symptoms, while only 22% remained unchanged. 11% of the patients died due to vasospasms. The authors concluded that, in comparison with the literature, the mortality in patients with angiographically confirmed vasospasms is significantly reduced.

Other CAs have been investigated in experimental SAH. Diltiazem displays cerebrovascular activity [140, 141, 168, 169] and was shown to ameliorate vasospasms and neurologic deficits in monkeys [170]. The inflammatory and infiltrative changes in the tunica media and tunica adventitia, which have been observed after SAH in monkeys, [171, 172] are reduced following pretreatment with diltiazem.

Since *rebleeding* is one of the major complicating factors in SAH, attempts have been made to counteract the increased risk of a second hemorrhage. Antifibrinolytic substances such as tranexamic acid and ε-aminocaproic acid have shown beneficial effects in small studies [173, 174] and even in a few larger, controlled studies [175, 176]. However, the mortality rate after 3 months of observation was unchanged, probably as a result of the increase in ischemic complications. It seems worthwhile, therefore, to explore the combined use of antifibrinolytics and CAs. Kassell et al. [177] used ε-aminocaproic acid and the CA nicardipine, which has been shown to influence significantly cerebrovascular smooth muscle tone [178–182]. Despite the fact that this study in 42 patients was uncontrolled, the mortality rate of 7% was clearly lower than that of the Cooperative Aneurysm Study [176] or the British-Dutch Study, both with antifibrinolysis alone. The incidence of ischemic complications in those studies was higher than in the aminocaproic acid-nicardipine protocol.

Cerebral Infarction

Cerebral infarction is due to focal, transient or permanent ischemia and does not affect the total brain. Complete and permanent ischemia of a brain area-limited in territorial size, may go along with survival of the brain and even with the lack of any clinical signs. Complete lack of blood perfusion to the total brain, however, is tolerated only for a very short time.

In this review, cerebral infarction and its tentative treatment by CAs is deliberately separated from the protection afforded by these substances against total brain ischemia. With respect to the specific effects of CAs in focal or total brain ischemia, the initial reversible sequestration of calcium must be separated from the irreversible Ca^{2+} overload associated with cell necrosis [183].

It was shown that nimodipine increases blood flow in experimentally induced focal ischemia [184–186], both in cats and monkeys. However, it may simultaneously impair autoregulation to reduced perfusion pressure, and altered reactivity of the vascular wall was shown in baboons [184]. Gelmers [187] showed in 3 out of 10 patients with acute ischemic stroke that the increase of CBF following administration of nimodipine was even more pronounced in the ischemic focus than in the normal tissue. This selective effect was not also in subacute infarction, head injuries and SAH with vasospasms [188].

Nimodipine can indeed cross the blood-brain barrier, but it is believed that its action on the cerebrovascular wall is more pronounced. This smaller effect on the neuronal cells might explain why nimodipine decreases the threshold for development of brain edema and electrolyte disturbances, and therefore favors the development of ischemic cell damage.

Other authors have shown that nimodipine (670 ng/kg/min i.v.) reduces vascular resistance behind a completely ligated middle cerebral artery. Due to reduction of systemic blood pressure, this vasodilatation did not result in an increase in CBF [189]. In gerbils, however, nimodipine (0.1 mg/kg) reduces mortality after bilateral occlusion of the common carotid artery and infarct size was reduced 1 h after occlusion of the middle cerebral artery [190–192].

Nowicki et al. [193] reported that diltiazem, nifedipine, perhexilene and verapamil are potent vasodilatators, but that they do not preserve the functional integrity of isolated cerebral mitochondria. This is of particular importance since cerebral ischemia results in a rapid decline of the mitochondrial ability to synthetize ATP. In addition, the remaining energized mitochondria may fail to sequester the increased Ca^{2+} accumulated in the ischemic tissue [194].

One controlled trial of nimodipine in acute ischemic stroke [195] concerns 186 patients, who were allocated to a prospective randomized, double-blind study. Therapy was started within 24 h after the onset of symptoms. The control group received low-molecular-weight dextran plus placebo instead of nimodipine, and the mortality in this group was 20.4% after 4 weeks. In the group treated with dextran and nimodipine, the mortality was 8.6%. However, this effect was restricted to male patients. The surviving patients in the nimodipine group (who received 120 mg nimodipine per day) improved more than the surviving patients in the control group. Patients with a moderate to severe deficit at the entrance into the study improved most of all.

The study describes such a remarkable beneficial effect of this CA that further multicenter studies should be performed as soon as possible. Other dihydropyridine compounds like nicardipine may affect cellular function in cerebral ischemia as well. At least, this was shown for the somatosensory-evoked potentials as a parameter reflecting brain tissue function [196]. Since nicardipine is taken up by ischemic brain tissue in the rat [197], it may be argued that the beneficial effect might partially be caused by direct metabolic action on the cells.

Verapamil was shown to inhibit the cellular uptake of Ca^{2+} ions in the hippocampus of the rabbit during ischemia and therefore indicates a direct effect on brain tissue cells as well [198].

The potential value of CAs in ameliorating the functional consequences of cerebral ischemia is probably not based on an increase in CBF alone [199]. Prostacyclin increases CBF in ischemic tissue of rats with middle cerebral artery occlusion without pH alteration [199]. Nimodipine, (0.5 mg/kg/min) and verapamil (40 mg/kg/min) did not change flow by correction to local tissue pH. The animals receiving the CA showed a tendency to develop smaller infarcts than the control animals. These and other [200] observations strongly suggest an additional metabolic effect in the brain.

With respect to focal ischemia, simulating a stroke-like model, the beneficial effect of CAs like flunarizine and nimodipine, given prior to ischemia [201], yield less important information. The fall in ATP in ischemia is retarded and glucose metabolism is improved by nimodipine pretreatment [202], and flunarizine pretreatment was shown to decrease the ischemic cell swelling [203]. However, since all CAs which act on the cerebrovascular system may affect normal vascular reactivity [184], it

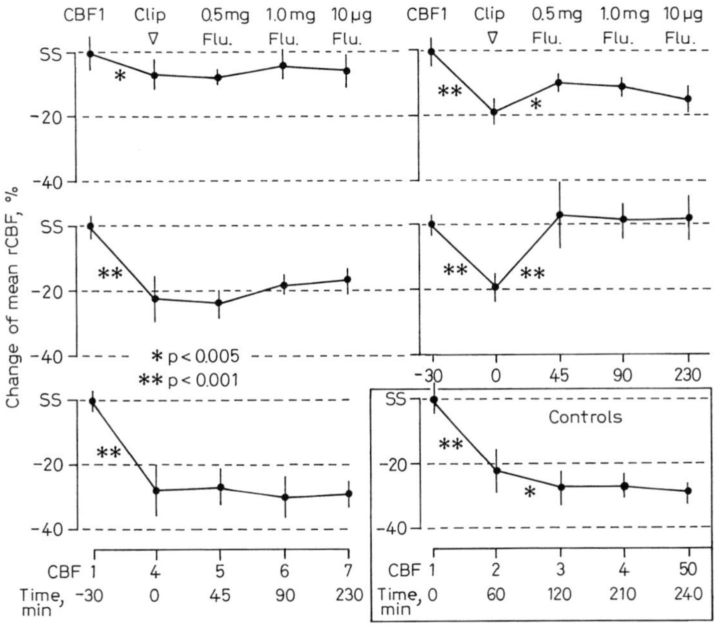

Fig. 3. Change of regional CBF (rCBF) in baboons with focal cerebral ischemia. rCBF is indicated as percentage change from steady-state value before ischemia on the ordinate. Abscissa: CBF 1–7, 30 min before induction of ischemia and up to 230 min after start of ischemia. The lower right curve indicates the mean value of 6 baboons with ischemia and no treatment (controls), whereas all other curves indicate the flow change in 5 individual baboons. It is obvious that in contrast to the CBF course in the controls the animals which were treated with flunarizine (Flu.) presented with significant interindividual changes in flow behavior. In only 2 of 5 animals was CBF increased after induction of ischemia and administration of flunarizine.

cannot be excluded that intracerebral steal phenomena with an increase of CBF in the normal and decrease of CBF (due to ischemia-induced vasoparalysis) in the ischemic tissue occur. This has been shown at least in some protocols [204]. Our own studies with flunarizine in normal baboons have indicated that the basic CBF is not affected, but that the physiological vasodilatation (following reduction of blood pressure) is impaired. In baboons with acute ischemia in the territory of the middle cerebral artery flunarizine leads to different effects with respect to tissue perfusion (fig. 3). Some baboons present with an increase of CBF, others with no reaction or even a further decrease in the hours following clipping of the middle cerebral artery, probably due to edema and enlargement of the penumbra. This partially negative effect in a model which aims at simulating the clinical situation is not surprising since many other secondary effects act on CBF after induction of middle cerebral artery ischemia (like development of both intra- and extracellular edema, cell necrosis with change of osmolarity, intravascular rheologic phenomena leading to further reduction of flow including stasis, development of tissue pressure gradients). Therefore, it would be surprising if administration of one (any) CA following occlusion of cerebral arteries restores cerebral blood flow in all subjects. However, these promising results should stimulate further experimental and clinical studies.

Calcium Antagonists and Global Cerebral Ischemia

Global cerebral ischemia is different from *focal* ischemia since several protecting mechanisms like rest perfusion via anastomotic networks, opening of collaterals, partial preservation of metabolism are not available. Cardiopulmonary arrest is one clinical correlate of total cerebral ischemia. Functional recovery from cardiopulmonary arrest, which affects the circulation of all organs, is determined to a major degree by the recovery of the neurons. During the arrest, both CBF and oxygen delivery decrease rapidly without any compensating mechanism. The patient looses consciousness within a few seconds. However, evidence has accumulated that the definite neurologic deficit is not due solely to immediate neuronal death. Recovery of high-energy metabolism, other metabolic steps and action potential generation have been shown after periods of total ischemic anoxia up to 60 min [205, 206].

The threshold for synaptic transmission deteriorates at an early point after arrest [207–210], the EEG flattens and evoked-potential responses cease. At this level, however, the concentration of ATP is still almost normal, but phosphocreatinine decreases, whereas lactate concentration increases [205]. If CBF collapses at the level of membrane function, lactate concentration increases further and K^+ leaks out of the cell [211–213]. The calcium concentration inside the cell starts to increase, then uncouples oxidative phosphorylation and activates the phospholipases [214], which results in release of fatty acids, mobilization of free radicals, thromboxanes and leukotrienes. After reperfusion, damage continues with cell swelling [215], intravascular stasis [216], both contributing to the no-reflow phenomenon [217]. Precapillary vasoconstriction provoked by Ca^{2+} activation further impairs tissue perfusion and results in the secondary (delayed) hypoperfusion syndrome [218, 219]. In the brain subjected to transient global ischemia, different degrees of cell damage may be observed simultaneously: some cells may be dying, others still preserve their substrate metabolism, but lack functional integrity, and others may be operating between the threshold of synaptic transmission and that of membrane function and therefore belong to the penumbra [25]. Particularly the latter cell group is capable of full recovery, but is functionally paralyzed and needs rapid normalization of blood flow and metabolism. Intracellular Ca^{2+} accumulation and Ca^{2+} overload in the mitochondria and dendrites of selectively vulnerable neurons [220, 221] should be reversed as early as possible to avoid definite cell damage. The structural damage in incomplete ischemia may be greater than in complete ischemia, since in the former condition glucose is supplied to a certain extent, resulting in continued production of lactate and development of intracellular acidosis.

After complete ischemia with reperfusion, the secondary processes lead to a state of delayed and prolonged hypoperfusion [222, 223]. This occurs at a time when the neurons are actually capable of functional recovery. Since cerebrovascular resistance increases in this stage without a rise in intracranial pressure, a cerebral vasodilator might restore normal flow and have some impact on the neurological outcome. However, the vasodilatation should affect the vessels of the diseased territory only and not those of normal tissue. Nimodipine in this connection has been shown to increase CBF in normal dogs [224] and in dogs subjected to a period of transient complete ischemia [225, 226]. In the normal dog, CBF increases to a greater extent in animals in which the blood-brain barrier was opened by the administration of urea prior to nimodipine, indicating that nimodipine probably has an effect on the pathologic tissue as well. The results have been confirmed in cats [227].

The postischemic hyperemia, which follows the release of complete ischemia very rapidly but only for a short time, is depressed by nimodipine, both upon pretreatment and also when administered after ischemia of 10 min duration. The delayed hypoperfusion following the hyperemic phase after about 15–30 min [228, 229] is partially neutralized by pre- and posttreatment. However, the neurologic improvement in the dogs was noted only upon pretreatment with nimodipine. These data are of particular importance, since the delayed hypoperfusion definitely contributes to the final outcome [230].

Another CA (flunarizine) has been shown to increase CBF and oxygen consumption after complete ischemia [231]. In rats and mice, flunarizine protects the brain against hypoxia [232]. However, Newberg et al. [233] did not find any beneficial effect of flunarizine (0.6 mg/kg) on CBF, $CMRO_2$, ATP, phosphocreatinine and behavior after 10 min of complete ischemia. This seems to be in agreement with the observation of Hossmann et al. [234], who could not observe an early influence on intracellular Ca^{2+} accumulation by flunarizine in cats. Neither biochemical nor electrical recovery was supported in these experiments. Van Reempts et al. [235] and Wauquier et al. [236], however, observed a restriction of neuronal damage by hypoxia due to nitrogen exposure after treatment with flunarizine (40 mg/kg).

Other CAs, like verapamil (0.1 mg/kg) and $MgSO_4$ (100 mg/kg), lead to preservation of CBF and oxygen consumption after resuscitation in dogs with cardiac arrest for 20 min [237]. Positive results with respect to the action of flunarizine have been published by Silverstein et al. [238]: in 7-day-old (immature) rats, neuronal damage was provoked by ligation of one internal carotid artery and exposure to 8% oxygen for 2 h. Morphologic alterations were observed in the hippocampus, striatum and cortex and neurotransmitter metabolism was changed as well [239]. Flunarizine (20 mg/kg) proved able to neutralize these changes partially. This neuroprotective effect could be related to an increase in CBF [240], or to the scavenging of free radicals, as shown for the myocardium [241]. Furthermore, it may be speculated that flunarizine has acted directly via specific recognition sites which have been proven to exist for this CA [242]. However, the functional importance of the sites remains unknown since in some studies calcium fluxes into synaptosomes due to depolarisation were not prevented by CA [243]. Other groups, however, have shown an impairment of neurotransmitter release [244] induced by nitrendipine [245].

An important action of CA on cerebral metabolism during global cerebral ischemia has been shown by Heffez und Passoneau [224] in Mongolian gerbils, subjected to bilateral clamping of both common carotid arteries for 1–5 min of recirculation: nimodipine (1 mg/kg) retarded the decrease of the ATP level in cortical and striatal tissues after 1 min of ischemia and at the same time the glucose concentration was improved.

In summary, it seems highly probable that intracellular Ca^{2+} accumulation plays one of the detrimental roles for functional and morphological recovery after transient complete cerebral ischemia. Although a functional benefit has not been proven for any CA when administered *after* the ischemic event, clinical implications might be possible in the light of cardiovascular and neurovascular surgery, which might lead to transient interruption of blood flow. However, even experimental studies have not been performed in sufficient number to clarify the possible influence of impairment of excessive Ca^{2+} entry into the intracellular space. Since at present no reliable concept exists for counteracting pathophysiologic complications of transient complete brain ischemia, further studies are necessary.

Vertebrobasilar Insuffiency and Vertigo

Vertebrobasilar insufficiency is a vague term. It implicates insufficient perfusion of cerebral tissue by the vertebral arteries. Clinical symptoms like transient double or blurred vision, dysarthric speech, unsteady gait or standing, hearing impairment, vertigo, drop attacks and other signs are indicators of this insufficient function. However, in practice, the diagnosis of vertebrobasilar insufficiency is made too often.

Vertigo, which has many potential different etiologies, has been treated experimentally in several studies with flunarizine, nimodipine and cinnarizine. In rabbits, flunarizine (10 mg/kg i.p.) reduces rotation-induced nystagmus [246]. The same effect was noted with flunarizine 5 mg/kg i.v. on caloric-induced nystagmus. Nimodipine (1 and 10 mg/kg) did not affect nystagmoid reactions after rotation but normalized the electro-nystagmogram in dogs if one vertebral artery was ligated.

Several studies have shown a beneficial effect on nystagmus in patients treated with flunarizine [247–250]. Similar effects have been described in patients with vertigo of labyrinth origin [251–254] treated for 4–8 weeks; in more than 3/4 of the patients, improvement of autonomic signs and reduction of tinnitus were observed within a few weeks. Most of the studies have been performed in comparison with placebo. In patients with Menière's disease, an improvement of hearing was recorded [255] after flunarizine treatment.

It was shown by two groups of investigators that flunarizine exerts a beneficial effect in patients in whom the diagnosis of vertebrobasilar insufficiency was established [250, 256]. The action of flunarizine in this respect might be caused partially by its vestibular depressant activity, resulting in the reduction of nystagmus. Since flunarizine possesses an antihistaminic activity as well (but much less than cinnarizine) [257] this may be a further explanation. Other antihistaminics with anticholinergic and sedative actions are potentially beneficial in vertigo as well [258, 259].

Epilepsy

Experimental models have indicated that evoked epileptiform potentials (by penicillin administration) at the cortical surface are accompanied by small-amplitude depolarization and followed by steep re- and hyperpolarization. The depolarization increases in amplitude, leading to burst activity of rectangular shape – the so-called paroxysmal depolarization shift [260–263]. It has been observed in all experimental seizure models [264–267] and represents either a 'giant excitatory' postsynaptic potential [268] or an 'endogenously generated' membrane potential fluctuation [269, 270].

The paroxysmal depolarization shift is preceded by enhanced excitatory postsynaptic potentials which might favor synchronization if a critical level is passed [271–273]. The spread of this seizure activity is further supported by extracellular potassium activity [274–276]. However, a column of epileptic paroxysmal depolarization shift is surrounded in a cylindrical fashion by a zone of discharge inhibition [277, 278] or – if a paroxysmal depolarization shift occurs only in superficial layers – by vertical inhibition [279–281]. Termination of seizure activity probably depends partially on developing cell hypoxia and lactic acidosis with metabolic exhaustion [282]. Other processes, like increased activity of outward current, may play an additional role for the determination of epileptic activity [283]; since Ca^{2+} influx into neurons contributes to development and substantiation of epileptic activity [284] and since the prevention of cellular Ca^{2+} overload may have anticonvulsant effects, CAs may possess a certain therapeutic value.

Phenytoin, barbiturates and probably carbamazepine impair Ca^{2+} uptake at nerve terminals which then, by inhibition of neurotransmitter release, limit the spread of seizures [285–287]. The diphenylalkylamines, cinnarizine and flunarizine, inhibit motor correlates of pentylenetetrazol- and electroshock-induced seizure activity in the hindleg of mice [288]. Seizure spread as provoked by electroshock, alcylglycine or metrazol administration is impaired by flunarizine, whereas bicuculline-provoked seizures are less affected [289, 290].

Nimodipine prevents seizure activity induced by pentylenetetrazol but not that following electroshock [291, 292], whereas verapamil has no effect, probably due to its limited capability to cross the blood-brain barrier. The capacity of flunarizine to impair seizures in experimental protocols is somewhere in the range of that observed for phenytoin and phenobarbitone. However, despite the promising experimental results little clinical experience exists. In a small group of 33 patients with mainly complex partial seizures a double-blind cross-over study over 6 months (two 3-month periods) revealed that flunarizine (10 mg) daily, added to the conventional individual medication, reduced seizure activity by more than 50% in 7 cases and by 25–50% in 11 patients. Blood levels of the basic medication were unaltered [293].

Thereafter, the same authors [293] completed an open trial in 52 therapy-resistant patients who suffered from complex partial seizures. Flunarizine (initial doses of 10–15 mg daily) which was increased to a maximum of 25 mg daily over a period of 15 months, or until the development of intolerable side effects. About a third of the patients reported a seizure reduction of 50% mainly if high doses were taken. 6 out of 47 patients experienced an increase in seizure frequency. Side effects were drowsiness and weight increase, as reported by 19 out of 52 patients. The authors have evaluated the effect of flunarizine on an open basis in their epileptic outpatient clinic and consider it as an efficient adjuvant for antiepileptic therapy. Besides the effect on seizure frequency in many patients, the aura was abolished, which, however, is a suppression of a possible warning sign. A survey of all relevant studies with flunarizine in the treatment of epilepsy was recently presented by Binnie [284].

In summary, further controlled studies on the antiepileptic activity of flunarizine seem to be necessary. However, for the time being, it cannot yet be recommended for the monotherapy of epileptic seizures.

Conclusions

Nimodipine, a dihydropyridine, and the diphenylethylamine, flunarizine, are the only CAs that have been studied – in most cases on a modest scale – as potential therapeutics in various neurological disorders. The treatment of migraine was dealt with in a separate chapter (see pp. 143–152). The mode of action of the two aforementioned CAs as therapeutics in various neurological diseases remains rather unclear and there is no conclusive evidence that the therapeutic efficacy is caused by calcium antagonism per se, although this seems very likely. Globally, two important mechanisms should be taken into account. Nimodipine is a CA with strong vasodilator activity in the cerebral region, although the drug appears to possess cytoprotective activity as well. Flunarizine owes its therapeutic activity rather to intracellular cytoprotective activity, since it is hardly a vasodilator. The cytoprotective activity of flunarizine becomes predominantly manifest in conditions of ischemia, which indeed play a role in migraine, cerebral ischemia, cerebral infarction and SAH. The influence of flunarizine on epilepsy is possibly based upon its inhibitory influence of the neuronal influx of calcium ions, which appears to be involved in epileptic activity.

The therapeutic efficacy of flunarizine in vertebrobasilar insufficiency, and possibly in other forms of vertigo, is poorly understood and not necessarily the result of calcium antagonism.

Whatever the explanation of the therapeutic effects of the CAs may be, it appears to be of great interest that some new therapeutic approaches are shaping up for certain neurological disorders which are notoriously difficult to manage. As discussed on pp. 143–152, flunarizine has acquired a widely established position in the treatment of several forms of migraine. Several subgroups of patients with SAH appear to benefit from treatment with nimodipine, although a final judgement can only be based upon the outcome of ongoing controlled trials. Flunarizine is investigated as a potential therapeutic agent in SAH as well, but the outcome is unknown so far. Both in cerebral infarction and in global cerebral ischemia, limited clinical trials suggest a beneficial effect of nimodipine, but more data are required before a general recommendation for the use of this drug is justified. Flunarizine, which is now available for intravenous administration, is being studied as a potential remedy in acute stroke. Both flunarizine and cinnarizine appear to be beneficial in vertebrobasilar insufficiency and possibly also in other forms of vertigo, but, as discussed above, it is uncertain whether the therapeutic efficacy is directly related to calcium antagonism per se.

Finally, flunarizine appears to be useful as add-on therapy in certain forms of epilepsy which prove to be refractory to treatment with classical antiepileptic drugs.

References

1 Siesjö, B.K.: Brain energy metabolism (Wiley, New York 1978).

2 Sokoloff, L.: Metabolism of the central nervous system in vivo; in Field, Magoun, Hall, Handbook of physiology, sect. 1: Neurophysiology, vol. 3, pp. 1843–1864 (American Physiological Society, Washington 1960).

3 McIlwain, H.: Biochemistry and the central nervous system; 2nd ed. (Little, Brown, Boston 1959).

4 Arnfred, I.; Secher, O.: Anoxia and barbiturates: tolerance to anoxia in mice influenced by barbiturates. Archs int Pharmacodyn. Ther. *139:* 67–74 (1962).

5 Myers, R.E.: A unitary theory of causation of anoxia and hypoxia brain pathology; in Fahn, Davis, Rowland, Advances in neurology, vol. 26, pp. 195–213 (Raven Press, New York 1979).

6 Pulsinelli, W.A.; Waldmann, S.; Rawlinson, D.; Plum, F.: Hyperglycemia converts ischemic neurological damage into brain infarction. Neurology *32:* 1239 (1982).

7 Salford, L.G.; Plum, F.; Siesjö, B.K.: Graded hypoxia-oligemia in rat brain. I. Biochemical alterations and their implications. Archs Neurol. *29:* 227 (1973).

8 Plum, F.: What causes infarction in ischemic brain? The Robert Wartenberg Lecture. Neurology *33:* 222 (1983).

9 Marshall, L.F.; Welsh, F.; Durity, F.; Lounsbury, R.; Graham, D.J.; Langfitt, T.W.: Experimental cerebral oligemia and ischemia produced by intracranial hypertension. 3. Brain energy metabolism. J. Neurosurg. *43:* 323 (1975).

10 Rehncrona, S.; Mela, L.; Siesjö, B.K.: Recovery of brain mitochrondrial function in the rat after complete and incomplete cerebral ischemia. Stroke *10:* 437 (1979).

11 Ljunggren, B.; Schutz, H.; Siesjö, B.K.: Changes in energy state and acid-base parameters of the rat brain during complete compression ischemia. Brain Res. *73:* 277 (1974).

12 Nordström, C.H.; Rehncrona, S.; Siesjö, B.K.: Restitution of cerebral energy state, as well as of glycolytic metabolites, citric and cycle intermediates and associated amino acids after 30 min. of complete ischemia in rats anesthetized with nitrous oxide or phenobarbital. J. Neurochem. *30:* 479–486.

13 Nordström, C.H.; Rehncrona, S.; Siesjö, B.K.: Effects of phenobarbital in cerebral ischemia. II. Restitution of cerebral energy state, as well as of glycolytic metabolites, citric acid cycle intermediates and associated amino acids after pronounced incomplete ischemia. Stroke *9:* 335 (1978b).

14 Miller, J.R.; Myers, R.E.: Neurological effects of systemic circulatory arrest in the monkey. Neurology *20:* 715 (1970).

15 Neely, W.A.; Youmans, J.R.: Anoxia of canine brain without damage. J. Am. med. Ass. *183:* 1085 (1963).

16 Hossmann, K.A.: Total ischemia of the brain; in Zülch, Kaufmann, Hossmann, Hossmann, Brain and heart infarct, pp. 107–122 (Springer, Berlin 1977).

17 Schmahl, F.W.; Betz, E.; Dettinger, E.; Hohorst, H.J.: Energiestoffwechsel der Grosshirnrinde und Elektroenzephalogramm bei Sauerstoffmagel. Pflügers Arch. *292:* 46 (1966).

18 Siesjö, B.K.; Nilsson, L.: The influence of arterial hypoxemia upon labile phosphates and upon extracellular and intracellular lactate and pyruvate concentrations in the rat brain. Scand. J. clin. Lab. *27:* 83 (1971).

19 Hossmann, K.A.; Lechtape-Gruter, H.; Hossmann, V.: The role of cerebral blood flow for the recovery of the brain after prolonged ischemia. Z. Neurol. *204:* 281 (1973).

20 Llinas, R.: The role of calcium in neuronal function; in Schmitt, Worden, The neurosciences: fourth study program, pp. 555–571 (MIT Press, Cambridge 1979).
21 Astrup, J.; Rehncrona, S.; Siesjö, B.K.: The increase in extracellular potassium concentration in the ischemic brain in relation to the preischemic functional activity and cerebral metabolic rat. Brain Res. *199:* 161 (1980).
22 Hansen, A.J.: Extracellular potassium concentration in juvenile and adult rat brain cortex during anoxie. Acta physiol. scand. *99:* 412 (1977).
23 Hossmann, K.A.; Sakaki, S.; Zimmermann, V.: Cation activities in reversible ischemia of the cat brain. Stroke *8:* 77 (1977).
24 Harris, R.J.; Symon, L.; Branston, M.; Bayhan, M.: Changes in extracellular calcium activity in cerebral ischemia. J. cereb. Blood Flow Metab. *1:* 203 (1981).
25 Astrup, J.; Siesjö, B.K.; Symon, L.: Thresholds in cerebral ischemia – the ischemic penumbra. Stroke *12:* 723–725 (1981).
26 Sharbrough, F.W.; Messick, J.M., Jr.; Sundt, T.M., Jr.: Correlation of continuous electroencephalograms with cerebral blood flow measurements during carotid endarterectomy. Stroke *4:* 674–683 (1973).
27 Sundt, T.M., Jr.; Sharbrough, F.W.; Piepgras, D.G.; Kearns, T.P.; Messick, J.M., Jr.; O'Fallon, W.M.: Correlation of cerebral blood flow and electroencephalographic changes during carotid endarterectomy with results of surgery and hemodynamics of cerebral ischemia. Mayo Clin. Proc. *56:* 533–543 (1981).
28 Hossmann, K.A.; Sato, K.: Recovery of neuronal function after prolonged cerebral ischemia. Science *168:* 375–376 (1970).
29 Hossmann, K.A.: Cortical steady potential, impedance and excitibility changes during and after total ischemia of cat brain. Expl. Neurol. *32:* 163–175 (1971).
30 Hossmann, K.A.; Sakaki, S.; Zimmermann, V.: Cation activities in reversible ischemia of the cat brain. Stroke *8:* 77–81 (1977b).
31 Strup, J.; Rehncrona, S.; Siesjö, B.K.: The increase in extracellular potassium concentration in the ischemic brain in relation to the preischemic functional activity and cerebral metabolic rat. Brain Res. *199:* 161–174 (1980).
32 Astrup, J.; Symon, L.; Branston, N.M.; Lassen, N.A.: Cortical evoked potential and extracellular K^+ and H^+ at critical levels of brain ischemia. Stroke *8:* 51–57 (1977).
33 Branston, N.M.; Strong, A.J.; Symon, L.: Impedance related to local blood flow in cerebral cortex. J. Physiol., Lond. *275:* 81–82 (1978).
34 Watanabe, O.; West, C.R.; Bremer, A.: Experimental regional cerebral ischemia in the middle cerebral artery territory in primates. 2. Effects on brain water and electrolytes in the early phase of MCA. Stroke *8:* 71–76 (1977).
35 Olsson, Y.; Crowell, R.M.; Klatzo, I.: The blood-brain barrier to protein tracers in focal cerebral ischemia and infarction caused by occlusion of the middle cerebral artery. Acta neuropath. *18:* 89–102 (1971).
36 Kamijyo, Y.; Garcia, J.H.; Cooper, J.: Temporary regional cerebral ischemia in the cat. A model of hemorrhagic and subcortical infarction. J. Neuropath. exp. Neurol. *36:* 338–350 (1977).
37 Ito, U.; Go, W.G.; Walker, J.T., Spatz, M.; Klatzo, I.: Experimental cerebral ischemia in Mongolian gerbils. 3. Behavior of the blood-brain barrier. Acta neuropath. *34:* 1–6 (1976).
38 Zimmermann, V.; Hossmann, V.; Hossmann, K.A.: Intracranial pressure after pro-

longed cerebral ischemia; in Lundberg, Ponten, Brock, Intracranial pressure, vol. 2, pp. 177–182 (Springer, Berlin 1975).

39 Ames, A. III; Wright, R. L.; Kowada, M.; Thurston, J. M.; Majno, G.: Cerebral ischemia. II. The no-reflow phenomenon. Am. J. Path. *52:* 437–453 (1968).

40 Kassel, N. F.; Sasaki, T.; Colohan, A. R. T.; Nayar, G.: Cerebral vasospasm following aneurysmal subarachnoid hemorrhage. Stroke *16:* 562–572 (1985).

41 Wellum, G. R.; Peterson, J. W.; Zervas, N. T.: The relevance of in vitro smooth muscle experiments to cerebral vasospasm. Stroke *16:* 573–581 (1985).

42 Fisher, C. M.; Roberson, G. H.; Ojemann, R. G.: Cerebral vasospasm with ruptured sacular aneurysm – the clinical manifestation. Neurosurgery *1:* 245–248 (1977).

43 Ferguson, G. G.; Harper, A. M.; Fitch, W.; et al.: Cerebral blood flow measurements after spontaneous subarachnoid hemorrhage. Eur. Neurol. *8:* 15–22 (1972).

44 Heinbrun, M. P.; Olsen, J.; Lassen, N. A.: Regional cerebral blood flow studies in subarachnoid hemorrhage. J. Neurosurg. *37:* 36–44 (1972).

45 Symon, L.; Ackerman, R.; Bull, J. W. D.; et al.: The use of the xenon clearance method in subarachnoid hemorrhage. Post-operative studies with clinical and angiography correlation. Eur. Neurol. *8:* 8–14 (1972).

46 Grubb, R. L.; Raichle, M. E.; Eichlin, J. O.; Gado, M. H.: Effects of subarachnoid hemorrhage on cerebral blood volume, blood flow and oxygen utilization in humans. J. Neurosurg. *46:* 446–453 (1977).

47 Pickard, J. D.; Boisvert, D. P. J.; Graham, D. J.; Fitch, W.: Late effects of subarachnoid hemorrhage on the response on the primate cerebral circulation to drug-induced changes in arterial blood pressure. J. Neurol. Neurosurg. Psychiat. *42:* 899–903 (1979).

48 Weir, B.; Rothberg, C.; Grace, M.; Davis, F.: Relative prognostic significance of vasospasm following subarachnoid hemorrhage. Can. J. neurol. Sci. *2:* 109–114 (1975).

49 Auer, L. M.: Acute surgery of cerebral aneurysms and prevention of symptomatic vasospasm. Acta neurochir. *69:* 273–281 (1983).

50 Ljunggren, B.; Säveland, H.; Brandt, L.: Causes of unfavorable outcome after early aneurysm operation. Neurosurgery *13:* 629–633 (1983).

51 Suzuki, J.; Kodama, N.; Yoshimoto, K.; Mizoi, K.: Ultraearly surgery of intracranial aneurysms. Acta neurochir. *63:* 186–191 (1982).

52 Suzuki, J.; Komatsu, S.; Sato, T.; Sakurai, Y.: Correlation between CT findings and subsequent development of cerebral infarction due to vasospasm in subarachnoid hemorrhage. Acta neurochir. *55:* 63–70 (1980).

53 Taneda, M.: Effect of early operation for ruptured aneurysms on prevention of delayed ischemic symptoms. J. Neurosurg. *57:* 622–627 (1982).

54 Gurusinghe, N. T.; Richardson, A. E.: The value of computerized tomography in aneurysmal subarachnoid hemorrhage. J. Neurosurg. *60:* 763–770 (1984).

55 Mitsukami, M.; Takemae, T.; Tayawa, T.; Kawase, T.; Matsuzaki, T.: Value of computed tomography in the prediction of cerebral vasospasm after aneurysm rupture. Neurosurgery *7:* 583–586 (1980).

56 Mitsukami, M.; Kawase, T.; Usami, T.; Tazawa, T.: Prevention of vasospasm by early operation with removal of subarachnoid blood. Neurosurgery *10:* 301–307 (1982).

57 Taneda, M.: Effect of early operation for ruptured aneurysm on prevention of delayed ischemic symptoms. J. Neurosurg. *57:* 622–628 (1982).

58 Ljunggren, B.; Brandt, L.; Saeveland, H.; Nilsson, P.E.; et al.: Outcome in 60 consecutive patients treated with early aneurysm operation and intravenous nimodipine. J. Neurosurg. *61:* 864–873 (1984).
59 Ljunggren, B.; Brandt, L.; Kagstroem, E.; Sundberg, G.: Results of early operations for ruptured aneurysms. J. Neurosurg. *54:* 473–479 (1981).
60 Suzuki, J.: Grading and timing of the operation on cerebral aneurysms; in Pia, Langmaid, Zierski, Cerebral aneurysms. Advances in diagnosis and therapy, pp. 203–208 (Springer, Berlin 1979).
61 Sano, K.; Saito, I.: Timing of microsurgical treatment of ruptured intracranial aneurysms; in: Wilkins, Cerebral arterial spasm, pp. 447–454 (Williams & Wilkins, Baltimore 1980).
62 Suzuki, J.: Cerebral vasospasm: prediction, prevention and protection; in Pia, Langmaid, Zierski (eds), Cerebral aneurysms. Advances in diagnosis and therapy, pp. 155–162 (Springer, Berlin 1979).
63 Osaka, K.: Prolonged vasospasm produced by breakdown products of erythrocytes. J. Neurosurg. *47:* 403–411 (1977).
64 Starling, L.M.; Boullin, D.J.; Grahame-Smith, D.G.; Adams, C.B.T.; Gye, R.S.: Responses of isolated human arteries to 5-hydroxytryptamine, noradrenaline, serum, platelets and erythrocytes. J. Neurosurg. Psychiat. *38:* 650–656 (1975).
65 Asano, T.; Tanishama, T.; Sasaki, T.; Sano, K.: Possible participation of free radical reactions initiated by clot lysis in the pathogenesis of vasospasm after subarachnoid hemorrhage; in Wilkins, Cerebral arterial spasm, pp. 190–201 (Williams & Wilkins, Baltimore 1980).
66 White, R.P.; Hagen, A.A.; Morgan, H.; Dawson, W.N.; Robertson, J.T.: Experimental study on the genesis of cerebral vasospasm. Stroke *6:* 52–57 (1975).
67 White, R.P.; Chapleau, C.E.; Dugdale, M.; Robertson, J.T.: Cerebral arterial contractions induced by human and bovine thrombin. Stroke *11:* 363–368 (1980).
68 Svendgaard, N.A.; Edvinson, L.; Olin, T.; Owman, C.; Sahlin, C.: On the pathophysiology of cerebral vasospasm: transmitter changes in perivascular sympathetic nerves and increased pial artery sensitivity to norepinephrine and serotonin; in Owman, Edvinccon, Neurogenic control of the brain circulation, pp. 143–152 (Pergamon Press, New York 1977).
69 Wilkins, R.H.: Attempted prevention or treatment of intracranial arterial spasm: a survey; in Wilkins, Cerebral arterial spasm, pp. 542–555 (Williams & Wilkins, Baltimore 1980).
70 Osaka, K.: Prolonged vasospasm produced by the breakdown products of erythrocytes. J. Neurosurg. *47:* 403–411 (1977).
71 Okwuasaba, F.; Cook, D.; Weir, B.: Changes in vasoactive properties of blood products with time and attempted identification of spasmogens. Stroke *12:* 775–780 (1981).
72 Toda, N.; Ozaki, T.; Ohta, T.: Cerebrovascular sensitivity to vasoconstricting agents induced by subarachnoid hemorrhage and vasospasm in dogs. J. Neurosurg. *46:* 296–303 (1977).
73 Tanishima, T.: Cerebral vasospasm: contractile activity of hemoglobin in isolated canine basilar arteries. J. Neurosurg. *53:* 787–793 (1980).
74 Barrows, L.J.; Hunter, F.T.; Banker, B.Q.: The nature and clinical significance of pigments in the cerebrospinal fluid. Brain *78:* 59–80 (1955).

75 Ozaki, N.; Mullan, S.: Possible role of the erythrocyte in causing prolonged cerebral vasospasm. J. Neurosurg. *51:* 773–778 (1979).
76 Wellum, G.R.; Irvine, T.W.; Zervas, N.T.: Dose responses of cerebral arteries of the dog, rabbit and man to human hemoglobin in vitro. J. Neurosurg. *53:* 486–490 (1980).
77 Allen, G.S.; Henderson, M.L.; Chou, S.N.; French, L.A.: Cerebral arterial spasm. I. In vitro contractile activity of vasoactive agents on canine basilar and middle cerebral arteries. J. Neurosurg. *40:* 433–441 (1974).
78 Allen, G.S.; Gross, C.J.; French, L.A.; Chou, S.N.: Cerebral arterial spasm. 5. In vitro contractile activity of vasoactive agents including human CSF on human basilar and anterior cerebral arteries. J. Neurosurg. *44:* 594–600 (1976).
79 Lobato, R.D.; Marin, J.; Salaices, M.; Rivilla, F.; Burgos, J.: Cerebrovascular reactivity to noradrenaline and serotonin following experimental subarachnoid hemorrhage. J. Neurosurg. *53:* 480–485 (1980).
80 Fraser, R.A.; Stein, B.M.; Barrett, R.E.: Noradrenergic meditation of experimental cerebrovascular spasm. Stroke *1:* 356–362 (1970).
81 Martins, A.N.; Doyle, T.F.; Wright, S.: Effect of topically applied serotonin on local cerebral blood flow. Stroke *9:* 25–28 (1978).
82 Endo, S.; Suzuki, J.: Experimental cerebral vasospasm after subarachnoid hemorrhage. Participation of adrenergic nerves in cerebral vessel wall. Stroke *10:* 703–711 (1979).
83 Edvinssion, L.; Egund, N.; Owman, C.H.; Sahlin, C.H.; Svendgaard, N.A.: Reduced noradrenaline uptake and retention in cerebrovascular nerves associated with angiographically visible vasoconstriction following experimental subarachnoid hemorrhage in rabbits. Brain Res. Bull. *9:* 799–805 (1982).
84 Pennink, M.; White, R.P.; Crockarell, J.R.; Robertson, J.T.: Role of prostaglandin, F_{2a} in the genesis of experimental cerebral vasospasm; angiographic and study in dogs. J. Neurosurg. *37:* 398–406 (1972).
85 Allen, G.S.; Gold, L.H.A.; Chou, S.N.; French, L.A.: Cerebral arterial spasm. 3. In vivo intracisternal production of spasm by serotonin and blood and its reversal by phenoxybenzamine. J. Neurosurg. *40:* 451–458 (1974).
86 Zervas, N.T.; Kuwayama, A.; Rosoff, C.B.; Salzmann, E.W.: Cerebral arterial spasm. Modification by inhibition of platelet function. Archs Neurol. *28:* 400–404 (1973).
87 Pasqualin, A.; da Pian, R.: An analysis of vasospasm following early surgery for intracranial aneurysms. Acta neurochir. *63:* 153–159 (1982).
88 Fisher, C.M.; Kistler, J.P.; Davis, J.M.: Relation of cerebral vasospasm to subarachnoid hemorrhage visualized by computerized tomographic scanning. Neurosurgery *6:* 1–9 (1980).
89 Gurusinghe, N.T.; Richardson, A.E.: The value of computerized tomography in aneurysmal subarachnoid hemorrhage. The concept of the CT score. J. Neurosurg. *60:* 763–770 (1984).
90 Bolton, T.B.: Mechanisms of action of transmitters and other substances on smooth muscle. Physiol. Rev. *59:* 606–715 (1979).
91 MacCalden, T.; Bevan, J.A.: Sources of activator calcium in rabbit basilar artery. Am. J. Physiol. *241:* H129–H133 (1981).
92 Ku, D.D.: Coronary vascular reactivity after acute myocardial ischemia. Science *218:* 576–578 (1982).

93 Zervas, N.; Hori, H.; Rosoff, C.: Experimental inhibition of serotonin by antibiotic: prevention of cerebral vasospasm. J. Neurosurg. *41:* 59–62 (1974).
94 Stemerman, M.B.; Baumgartner, H.R.; Spaet, T.H.: Subendothelial mircofibril and platelet adhesion. Lab. Invest. *24:* 175–186 (1971).
95 Ashford, T.P.; Freiman, D.G.: Platelet aggregation at sites of minimal endothelial injury. An electron mircoscopic study. Am. J. Path. *53:* 599–607 (1968).
96 Asano, T.; Tanishima, T.; Sasaki, T.; Sano, K.: Possible participation of free radical reactions initiated by clot lysis in the pathogenesis of vasospasm after subarachnoid hemorrhage; in Wilkins, Cerebral arterial spasm, pp. 190–201 (Williams & Wilkins, Baltimore 1980).
97 Sasaki, T.; Wakai, S.; Asano, T.; Watanabe, T.; Kirino, T.; Sano, K.: The effect of a lipid hydroperoxide of arachidonic acid on the canine basilar artery. J. Neurosurg. *54:* 357–365 (1981).
98 Chayatte, D.; Rusch, N.; Sundt, T.M.: Prevention of chronic experimental cerebral vasospasm with ibuprofen and high-dose methylprednisolone. J. Neurosurg. *59:* 925–932 (1983).
99 White, R.P.; Robertson, J.T.: Comparison of piroxicam, meclofenamate, ibuprofen, aspirin and prostacyclin efficacy in a chronic model of cerebral vasospasm. Neurosurgery 12: 40–46 (1983).
100 Fein, J.M.: Brain energetics and circulatory control after subarachnoid hemorrhage. J. Neurosurg. *45:* 498–507 (1978).
101 van Bremen, C.: Transmembrane calcium transport in vascular smooth muscle; in Bevan et al., Vascular neuroeffector mechanisms, pp. 67–69 (Karger, Basel 1975).
102 Allen, G.S.; Baughart, S.B.: In vitro effects of nifedipine on serotonin, phenylephrine and potassium-induced contractions of canine basilar and femoral arteries. 9. Neurosurgery *4:* 37–42 (1979).
103 Shimizu, K.; Ohta, T.; Toda, N.: Evidence for greater susceptibility of isolated dog cerebral arteries to Ca antagonists than peripheral arteries. Stroke *11:* 261–266 (1980).
104 Brandt, L.; Andersson, K.E.; Edvinsson, L.; Ljunggren, B.: Effects of extracellular calcium and of calcium antagonists on the contractile response of isolated human pial and mesenteric arteries; Thesis Lund (1981).
105 Allen, G.S.; Langhart, S.B.: Cerebral arterial spasm. 9. In vitro effects of nifedipine or serotonin, phenylephrine and potassium-induced contractions of canine basilar and femoral arteries. Neurosurgery *4:* 37–42 (1979).
106 Brandt, L.; Andersson, K.E.; Bengtsson, B.; et al.: An in vivo study. Surg. Neurol. *12:* 349–352 (1979).
107 Kazda, S.; Hoffmeister, F.; Garthoff, B.; Towart, R.: Prevention of the post-ischemic impaired reperfusion of the brain by nimodipine (Bay e 9736). Acta neurol. scand. *60:* suppl. 72, pp. 302–303 (1979).
108 Kazda, S.; Hoffmeister, F.: Effect of some cerebral vasodilators on the postischemic impaired cerebral reperfusion in cats. Archs Pharmacol., suppl. 307, R43 (1979).
109 Towart, R.; Perzborn, E.: Nimodipine inhibits carbocyclic thromboxane-induced contraction of cerebral arteries. Eur. J. Pharmacol. *69:* 213–215 (1981).
110 Towart, R.; Kazda, S.: The cellular mechanism of action of nimodipine (Bay e 9736), a new calcium antagonist. Br. J. Pharmacol. *67:* 409–410 (1979).

111 Auer, L.M.: Pial arterial vasodilatation by intravenous nimodipine in cats. Drug Res. *31:* 1423–1425 (1981).

112 Sahlin, C.; Brismar, J.; Ljunggren, K.; Owman, C.; Rodacki, M.A.; Salford, L.G.; Svendgaard, N.A.: The influence of late arterial spasm on CBF and its response to a Ca uptake inhibitor. Proc. 31st Annu. Meet. Nord. Neurokir. Fören. (Scand. Neurosurg. Soc.), Oulu 1979. Acta neurochir. *51:* 126–129 (1979).

113 Heffez, D.S.; Nowak, T.S.; Passonneau, J.V.: Nimodipine levels in gerbil brain following parenteral drug administration. J. Neurosurg. *63:* 589–592 (1985).

114 Peroutka, S.J.; Allen, G.S.: Calcium channel antagonist binding sites labeled by ^{3}H-nimodipine in human brain. J. Neurosurg. *59:* 933–937 (1983).

115 Brandt, L.; Anderson, K.E.; Edvinsson, L.; Ljunggren, B.: Effects of extracellular calcium and of calcium antagonists on the contractile responses of isolated human pial and mesenteric arteries. J. cereb. Blood Flow Metab. *1:* 339–347 (1981).

116 Towart, R.: The selective inhibition of serotonin-induced contractions of rabbit cerebral vascular smooth muscle by calcium antagonist dihydropyridines. Circulation Res. *48:* 650–657 (1981).

117 Towart, R.; Peryborn, E.: Nimodipine inhibits carbocylic thromboxane-induced contractions of cerebral arteries. Eur. J. Pharmacol. *69:* 213–215 (1981).

118 White, R.P.; Cunningham, M.P.; Robertson, J.I.: Effect of the calcium antagonist nimodipine on contractile responses of isolated canine basilar arteries induced by serotonin, prostaglandin F_2, thrombin and whole blood. Neurosurgery *10:* 344–348 (1982).

119 Cohen, J.C.; Allen, G.A.: Cerebral arterial spasm: the role of calcium in in vitro and in vivo analysis of treatment with nifedipine and nimodipine; in Wilkins, Cerebral arterial spasm, pp. 527–532 (William & Wilkins, Baltimore 1980).

120 Tanaka, K.; Gotoh, F.; Muramatsu, F.; Fukuuchi, Y.; Okayasu, H.; Suzuki, N.; Kobari, K.: Effect of nimodipine, a calcium antagonist, on cerebral vasospasm after subarachnoid hemorrhage in cats. Arzneimittelforsch. *32:* 1529–1534 (1982).

121 Kamiy, K.; Kuyama, H.; Symon, L.: An experimental study of the acute stage of subarachnoid hemorrhage. J. Neurosurg. *59:* 917–924 (1983).

122 Allen, G.S.; Bahr, A.L.: Cerebral arterial spasm. 10. Reversal of acute and chronic spasm in dogs with orally administered nifedipine. Neurosurgery *4:* 43–46 (1979).

123 Harper, A.M.; Craigen, L.; Kazda, S.: Effect of the calcium antagonist, nimodipine, on cerebral blood flow and metabolism in the primate. J. cerebr. Blood Flow Metab. *1:* 349–356 (1981).

124 Sahlin, C.; Delgado, T.; Owman, C.; Salford, L.G.; Svendgaard, N.-A.: Changes in cerebral blood flow and metabolism following intraarterial or local administration of nimodipine, before and after experimental subarachnoid hemorrhage in baboons. Stroke *17:* 220–224 (1986).

125 Kazda, S.; Garhoff, B.; Krause, H.P.; Schlossmann, K.: Cerebrovascular effects of the calcium antagonistic dihydropyridine derivate nimodipine in animal experiments. Drug Res. *32:* 331–338 (1982).

126 Harper, A.M.; Craigen, L.; Kazda, S.: Effect of the calcium antagonist, nimodipine, in cerebral blood flow and metabolism in the primate. J. cereb. Blood Flow Metab. *1:* 349–356 (1981).

127 Gaab, M.R.; Haubity, I.; Brawanski, A.; Korn, A.; Czech, T.: Acute effects of nimodipine on the cerebral blood flow and intracranial pressure. Neurochirurgica *28:* 93–99 (1985).

128 Svendgaard, N. A.; Brismar, J.; Delgado, T.; Egund, N.; Owman, C.; Rodacki, M. A.; Salin, C.; Salford, L. G.: Late cerebral arterial spasm: the cerebrovascular response to hypercapnia, induced by hypertension an the effect of nimodipine on blood flow autoregulation in experimental subarachnoid hemorrhage. Gen. Pharmacol. *14:* 167–172 (1983).

129 Gaab, M. R.; Höllerhage, H. G.; Zumkeller, M.; Trost, H. A.: The effect of the Ca antagonist nimodipine on cerebral blood flow autoregulation. J. cereb. Blood Flow Metab. *7:* suppl. 1, p. 170 (1987).

130 Brawley, B. W.; Strandness, D. E. Jr.; Kelly, W. A.: The biphasic response of cerebral vasospasm in subarachnoid hemorrhage. J. Neurosurg. *28:* 1–8 (1968).

131 Weir, B.; Grace, M.; Hansen, J.; Rothberg, C.: Time course of vasospasm in man. J. Neurosurg. *48:* 173–178 (1978).

132 Nosko, M.; Weir, B.; Krüger, C.; Cook, D.; Norris, S.; Overton, T.; Boisvert, D.: Nimodipine and chronic vasospasm in monkeys. I. Clinical and radiological findings. Neurosurgery *16:* 129–136 (1985).

133 Varsos, V. G.; Liszczak, T. M.; Han, D. H.; Kistler, J. P.; Vielma, J.; Black, P. M.; Heros, R. C.; Zervas, N. T.: Delayed cerebral vasospasm is not reversible by aminophylline, nifedipine or papaverine in a 'two-hemorrhage' canine model. J. Neurosurg. *58:* 11–17 (1983).

134 Heilbrun, M. P.; Olesen, J.; Lassen, N. A.: Regional cerebral blood flow studies in subarachnoid hemorrhage. J. Neurosurg. *37:* 36–44 (1972).

135 Wilkins, R. H.: Attempts at prevention or treatment of intracranial arterial spasm: an update. Neurosurgery *18:* 808–825 (1980).

136 Fleischer, A. S.; Raggio, J. F.; Tindall, G. T.: Aminophylline and isoproterenol in the treatment of cerebral vasospasm. Surg. Neurol. *8:* 117–121 (1977).

137 Norwood, C. W.; Poole, G. J.; Moody, D.: Treatment of experimental delayed cerebral arterial spasm with a beta-2-adrenergic stimulator and a phosphodiesterase inhibitor. J. Neurosurg. *45:* 491–497 (1976).

138 Auer, L. M.: Cerebrovascular activity of the calcium antagonist nimodipine; in Reivich, Hurtig, Cerebrovascular diseases, pp. 375–384 (Raven Press, New York 1983).

139 Kostron, H.; Twerdy, K.; Stampfl, G.; Mohsenipour, I.; Fischer, J.; Grunert, V.: Treatment of the traumatic cerebral vasospasm with the calcium channel blocker nimodipine: a preliminary report. Neurol. Res. *6:* 29–32 (1984).

140 Murata, S.; Nagao, T.; Nakajima, H.: Cerebral vasodilation and spasmolytic activity of diltiazem in anesthetized animals. Jap. J. Pharmacol. *32:* 1033–1040 (1982).

141 Bevan, J. A.: Selective action of diltiazem on cerebral vascular smooth muscle in the rabbit: antagonism of extrinsic but not intrinsic maintains tone. Am. J. Cardiol. *49:* 519–524 (1982).

142 McCalden, T. A.; Bevan, J. A.: The effect of calcium withdrawal and calcium antagonists on cerebrovascular tone and responses to various agonists; in Heistad, Marcus, Cerebral blood flow. Effects of nerves and neurotransmitters, pp. 21–27 (Elsevier Biomedical, New York 1982).

143 Shimizu, K.; Ohta, T.; Toda, N.: Evidence for greater susceptibility of isolated dog cerebral arteries to Ca agonists than peripheral arteries. Stroke *11:* 261–266 (1980).

144 Yamamoto, M.; Ohta, T.; Toda, N.: Mechanisms of relaxant action of nicardipine, a new Ca^{++} antagonist on isolated dog cerebral and mesenteric arteries. Stroke *14:* 270–275 (1983).

145 Shirasawa, Y.; White, R.P.; Robertson, J.T.: Mechanism of the contractile effect induced by uridine 5-triphosphate in canine cerebral arteries. Stroke *14:* 347–355 (1983).
146 Towart, R.: The selective inhibition of serotonin-induced contractions of rabbit cerebral vasular smooth muscle by calcium-antagonistic dihydropyridines. An investigation of the mechanism of action of nimodipine. Circulation Res. *48:* 650–657 (1981).
147 Allen, G.S.; Banghart, S.B.: Cerebral arterial spasm. 9. In vitro effects of nifedipine on serotonin-, phenylephrine- and potassium-induced contractions of canine basilar and femoral arteries. Neurosurgery *4:* 37–41 (1979).
148 Brandt, L.; Anderson, K.E.; Edvinssion, L.; Ljunggren, B.: Effects of extracellular calcium and of calcium antagonists on the contractile responses of isolated human pial and mesenteric arteries. J. cereb. Blood Flow Metab. *1:* 339–345 (1981).
149 Van Neuten, J.M.; Vanhoutte, P.M.: Calcium entry blockers and vasospasm; in Vanhoutte, Leusen, Cerebral vasodilatation, pp. 459–468 (Raven Press, New York 1981).
150 Borgers, M.; Thone, F.; van Reempts, J.; Verheyen, F.: The role of calcium in cellular dysfunction. Am. J. Emerg. Med. *1:* 154–161 (1983).
151 Allen, G.S.; et al.: Cerebral arterial spasm – a controlled trial of nimodipine in patients with subarachnoid hemorrhage. New Engl. J. Med. *308:* 619–624 (1983).
152 Post, K.; Flamm, E.S.; Goodgold, A.; Ransohoff, J.: Ruptured intracranial aneurysms. Case morbidity and mortality. J. Neurosurg. *46:* 290–295 (1977).
153 Saito, L.; Ueda, Y.; Sano, K.: Significance of vasospasm in the treatment of ruptured intracranial aneurysms. J. Neurosurg. *47:* 412–429 (1977).
154 Bewermeyer, H.; Szelies, B.; Lumenta, C.; Heiss, W.D.: Änderungen im Verlauf nach spontanen Subrachnoidalblutungen aus neurologischer Sicht. Fortschr. Neurol. Psychiat. *50:* 247–257 (1982).
155 Harries, A.D.: Subarachnoid hemorrhage and the electrocardiogram – a review. Postgrad. med. J. *57:* 294–296 (1981).
156 Doshi, R.; Neil-Dwyer, G.: A clinicopathological study of patients following a subarachnoid hemorrhage. J. Neurosurg. *52:* 295–301 (1980).
157 Cropp, G.J.; Manning, G.W.: Electrocardiographic changes in stimulating myocardial ischemia and infarction associated with spontaneous intracranial hemorrhage. Circulation *22:* 25–38 (1960).
158 Seiler, R.W.; Grolimund, P.; Zubruegg, H.R.: Evaluation of calcium-antagonist nimodipine for the prevention of vasospasm after aneurysmal subarachnoid haemorrhage. Acta neurochir. *85:* 7–16 (1987).
159 Seiler, R.W.; Grolimund, P.; Aaslind, R.; Huber, P.; Nornes, H.: Cerebral vasospasm evaluated by transcranial ultrasound correlated with clinical grade and CT-visualized subarachnoid haemorrhage. J. Neurosurg. *64:* 594–600 (1986).
160 Gilsbach, J.M.; Harders, A.: Early aneurysm operation and vasospasm – intracranial Doppler findings. Neurochirurgia *28:* 100–102 (1985).
161 Harris, R.J.; Branston, N.M.; Symon, L.; Bayhan, M.; Watson, A.: The effects of a calcium antagonist, nimodipine, upon physiological responses of the cerebral vasculature and its possible influence upon focal cerebral ischemia. Stroke *13:* 759–766 (1982).
162 Auer, L.M.; Oberbauer, R.W.; Schalk, H.V.: Human pial vascular reactions to intravenous nimodipine infusion during EC-IC bypass surgery. Stroke *14:* 210–213 (1983).

163 Steen, P.A.; Newberg, L.A.; Milde, J.H.; Michenfelder, J.D.: Nimodipine improves cerebral blood flow and neurologic recovery after complete cerebral ischemia in the dog. J. cereb. Blood Flow Metab. *3:* 38–43 (1983).
164 Dale, J.; Landmark, K.H.; Myhre, E.: The effect of nifedipine, a calcium antagonist, on platelet function. Am. Heart J. *105:* 103–105 (1983).
165 Vinge, E.; Andersson, K.E.; Brandt, L.; Ljunggren, B.; Nilsson, L.G.; Rosendal-Helgessen, S.; Pharmacokinetics of nimodipine in patients with aneurysmal subarachnoid hemorrhage. Eur. J. clin. Pharmacol. *30:* 421–425 (1986).
166 Böker, D.K.; Solymosi, L.; Wassmann, H.: Immediate postangiographic intraarterial treatment of cerebral vasospasm after subarachnoid hemorrhage with nimodipine. Neurochirurgica *28:* 118–120 (1985).
167 Koos, W.T.; Pernezky, A.; Auer, L.M.; Böker, D.B.; Gaab, M.R.; Jaschke, H.; Kostron, H.; Meinig, G.; Muizelaar, J.P.; van der Werft, A.J.M.; Seibert, H.K.; Ulrich, F.; Sprung, C.: Nimodipine treatment of ischemic neurological deficits due to cerebra vasospasms after subarachnoid hemorrhage. Neurochirurgica *28:* 114–117 (1985).
168 Sato, M.; Nagao, T.; Yamaguchi, I.; Nakajima, H.; Kiyomoto, A.: Pharmacological studies on a new 1,5-benzothiazepine derivative (CRD-401). I. Cardiovascular actions. Arzneimittel-Forsch. *21:* 1338–1343 (1971).
169 Nagao, T.; Murata, S.; Ikezawak, K.; Ikeo, T.; Narita, H.; Sato, M.: Effects of diltiazem on hemodynamics and His bundle electrogram in the anesthetized dog. Nippon Yakurigaku Zasshi *77:* 195–203 (1981).
170 Frazee, J.G.; Bevan, J.A.; Bevan, R.D.; Jones, K.R.: Effect of diltiazem on experimental chronic cerebral vasospasm in the monkey. J. Neurosurg. *62:* 912–917 (1985).
171 Bevan, J.A.; Bevan, R.D.; Frazee, J.G.: Chronic cerebrovascular spasm in the monkey; protection by diltiazem; in Bevan, Goldfraind, Maxwell, Stoclet, Worcel, Vascular neuroeffector mechanisms. Proc. 1st Int. Congr. Vascular Neuroeffector Mechanisms, Paris 1984, pp. 211–217 (Elsevier, Amsterdam 1985).
172 Bevan, J.A.; Bevan, R.D.; Frazee, J.G.: Functional arterial changes in chronic cerebrovasospasm in monkeys: an in vitro assessment of the contribution to arterial narrowing. Stroke *18:* 472–481 (1987).
173 Ramirez-Lassepas, M.: Antifibrinolytic therapy in subarachnoid hemorrhage caused by ruptured intracranial aneurysm. Neurology *31:* 316–322 (1981).
174 Adams, H.P. Jr.; Nibbelink, D.W.; Torner, J.C.; Sahs, A.L.: Antifibrinolytic therapy in patients with aneurysmal subarachnoid hemorrhage. Archs Neurol. *38:* 25–29 (1981).
175 Vermeulen, M.; Lindsay, K.W.; Murray, G.D.; Gheah, F.; Hijdra, A.; Muizelaar, J.P.; Schannong, M.; Teasdale, G.M.; van Grevel, H.; van Gijn, J.: Antifibrinolytic treatment in subarachnoid hemorrhage. New Engl. J. Med. *311:* 432–437 (1984).
176 Kassell, N.F.; Torner, J.C.; Adams, H.P.: Antifibrinolytic therapy in the acute period following aneurysmal subarachnoid hemorrhage. Preliminary observations from the Cooperative Aneurysm Study. J. Neurosurg. *61:* 225–230 (1984).
177 Kassell, N.F.; Torner, J.C.; Adams, H.P.: Antifibrinolytic therapy in the acute period following aneurysmal subarachnoid hemorrhage. J. Neurosurg. *61:* 562–572 (1984).
178 Peroutka, S.J.; Allen, G.S.: Calcium-channel antagonist binding sites labeled by ^{3}H-nimodipine in human brain. J. Neurosurg. *59:* 933–937 (1983).
179 Wilkins, R.H.: Attempts at prevention or treatment of intracranial arterial spasm. An update. Neurosurgery *18:* 808–825 (1986).

180 Alps, B.J.; Calder, C.; Wilson, A.: The beneficial effect of nicardipine compared with nifedipine and verapamil in limiting myocardial infarct size in baboons. Arzneimittel-Forsch. *33:* 868–876 (1983).

181 Fujiwara, S.; Duriyama, H.: Nicardipine actions on smooth muscle cells and neuromuscular transmission in the guinea pig basilar artery. J. pharmac. exp. Ther. *225:* 447–455 (1983).

182 Grotta, J.; Spydell, J.; Pettigrew, C.; Ostrow, P.; Hunter, D.: The effect of nicardipine on neuronal function following ischemia. Stroke *17:* 213–219 (1986).

183 van Reempts, J.; Borgers, M.: Ischemic brain injury and cell calcium: morphologic and therapeutic aspects. Ann. Emerg. Med. *14:* 736–742 (1985).

184 Harris, R.J.; Branston, N.M.; Symon, L.; Bayhan, M.; Watson, A.: The effects of a calcium antagonist, nimodipine, upon physiological responses of the cerebral vasculature and its possible influence upon focal cerebral ischemia. Stroke *13:* 759–766 (1982).

185 Hoffmeister, F.; Benz, U.; Heise, A.; Krause, H.P.; Neuser, V.: Behavioral effects of nimodipine in animals. Arzneimittel-Forsch. *32:* 347–360 (1982).

186 Kazda, S.; Garthoff, B.; Krause, H.P.; Schlossmann, K.: Cerebrovascular effects of the calcium antagonistic dihydropyridine derivative nimodipine in animal experiments. Arzneimittel-Forsch. *32:* 331–338 (1982).

187 Gelmers, H.J.: Effect of nimodipine on postischemic cerebrovascular reactivity, as revealed by mesuring regional cerebral blood flow (rCBF). Acta neurochir. *63:* 283–290 (1982).

188 Gaab, M.R.; Rode, C.P.; Schakel, E.H.; Haubitz, I.; Bockhorn, J.; Brawanski, A.: Zum Einfluss des Ca-Antagonisten Nimodipin auf die globale und regionale Hirndurchblutung. Klin. Wschr. *63:* 8–15 (1985).

189 Date, H.; Hossmann, K.A.: Effect of vasodilatating drugs on intracortical and extracortical vascular resistance following middle cerebral artery occlusion in cats. Ann. Neurol. *16:* 330–336 (1984).

190 Germano, I.M.; Bartkowski, H.M.; Cassel, M.E.; Pitts, L.H.: The therapeutic value of nimodipine in experimental focal cerebral ischemia. Neurological outcome and histopathological findings. J. Neurosurg. *67:* 81–87 (1987).

191 Sauter, A.; Rudin, M.: Calcium antagonists reduce the extent of infarction in rat middle cerebral artery occlusion model as determined by quantitative magnetic resonance imaging. Stroke *17:* 1228–1234 (1986).

192 Fujisawa, A.; Matsumoto, M.; Matsuyama, T.; Ueda, H.; Wanaka, A.; Yoneda, S.; Kimura, K.; Kamada, T.: The effect of the calcium antagonist nimodipine on the gerbil model of experimental cerebral ischemia. Stroke *17:* 748–752 (1986).

193 Nowicki, J.P.; MacKenzie, E.T.; Young, A.R.: Brain ischemia, calcium and calcium antagonists. Path. Biol. *30:* 282–288 (1982).

194 Fiskum, G.: Mitochondrial damage during cerebralischemia. Ann. Emerg. Med. *14:* 810–815 (1985).

195 Gelmers, H.J.; Gorter, K.; de Weerdt, C.J.; Wiezer, H.J.: A controlled trial of nimodipine in acute ischemic stroke. New Engl. J. Med. *318:* 203–207 (1988).

196 Grotta, J.; Spydell, J.; Pettigrew, C.; Ostrow, P.; Hunter, D.: The effect of nicardipine on neuronal function following ischemia. Stroke *17:* 213–219 (1986).

197 Grotta, J.C.; Pettigrew, L.C.; Lockwood, A.H.; Reid, C.: Brain extraction of a calcium channel blocker. Ann. Neurol. *21:* 171–175 (1987).

198 Hagberg, H.; Lehmann, A.; Hamberger, A.: Inhibition by verapamil of ischemic Ca^{2+} uptake in the rabbit hippocampus. J. cereb. Blood Flow Metab. *4:* 297–300 (1984).

199 Hakim, A.M.: Cerebral acidosis in focal ischemia. II. Nimodipine and verapamil normalize cerebral pH following middle cerebral artery occlusion in the rat. J. cereb. Blood Flow Metab. *6:* 676–683 (1986).

200 Heffez, D.S.; Passonneau, J.V.: Effect of nimodipine on cerebral metabolism during ischemia and recirculation on the Mongolian gerbil. J. cereb. Blood Flow Metab. *5:* 523–528 (1985).

201 Safar, P.: Recent advances in cardiopulmonary-cerebral resuscitation: a review. Ann. Emerg. Med. *13:* 856–862 (1984).

202 Young, W.L.; Josovitz, K.; Morales, O.; Chien, S.: The effect of nimodipine on post-ischemic cerebral glucose utilization and blood flow in the rat. Anesthesiology *67:* 54–59 (1987).

203 van Reempts, J.; Borgers, M.; van Deal, L.; van Eyndhoven, J.; van de Ven, M.: Protection with flunarizine against hypoxic-ischemic damage of the rat cerebral cortex. A quantitative morphologic assessment. Archs int. Pharmacodyn. Ther. *262:* 76–88 (1983).

204 Vorstrup, S.; Andersen, A.; Blegvad, N.; Paulson, O.B.: Calcium antagonist treatment may further decrease flow in ischemic areas in acute stroke. J. cereb. Flow Metab. *6:* 222–229 (1986).

205 Ames, A. III; Gurian, B.S.: Effects of glucose deprivation on function of isolated mammalian retina. J. Neurophysiol. *26:* 617 (1963).

206 Hossmann, K.A.; Kleihus, P.: Reversibility of ischemic brain damage. Archs Neurol. *29:* 375 (1973).

207 Heiss, W.D.: Flow threshold and morphological damage of brain tissue. Stroke *14:* 329–331 (1983).

208 Heiss, W.D.; Rosner, G.: Functional recovery of cortical neurons as related to degree and duration of ischemia. Ann. Neurol. *14:* 294–301 (1983).

209 Duffy, T.E.; Pulsinelli, W.A.: Regional cerebral glucose metabolism during hypoxia. Adv. Neurol. *26:* 287–292 (1979).

210 Astrup, J.; Siesjö, B.K.; Symon, L.: Threshold in cerebral ischemia – the ischemic penumbra. Stroke *12:* 723–725 (1981).

211 Astrup, J.: Energy-requiring cell functions in the ischemic brain: their critical supply and possible inhibition in protective therapy. J. Neurosurg. *56:* 482–497 (1982).

212 Siesjö, B.K.; Berntman, L.; Rehncrona, S.: Effect of hypoxia on blood flow and metabolism flux in the brain. Adv. Neurol. *26:* 267–283 (1979).

213 Rehncrona, S.; Rosen, I.; Siesjö, B.K.: Excessive cellular acidosis: an important mechanism of neuronal damage in the brain? Acta physiol. scand. *110:* 435–437 (1980).

214 Siesjö, B.K.: Brain energy metabolism (Wiley, New York 1978).

215 Little, J.R.: Microvascular alterations and edema in focal ischemia; in Pappius, Feindel, Dynamics of brain edema, pp. 237–243 (Springer, Berlin 1976).

216 Ginsberg, M.D.; Budd, W.W.; Welsh, F.A.: Diffuse cerebral ischemia in the cat. I. Local blood flow during severe ischemia and recirculation. Ann. Neurol. *3:* 482–492 (1978).

217 Ames, A.III; Wright, R.L.; Kowada, M.; Thurston, J.M.; Majno, G.: Cerebral ischemia. III. The no-reflow phenomenon. Am. J. Path. *52:* 437–453 (1986).

218 Gadzinski, D.S.; White, J.D.; Hoehner, P.J.; Hoehner, T.; Krome, C.; White, J.D.: Canine cerebral cortical blood flow and vascular resistance post cardiac arrest. Ann. Emerg. Med. *11:* 58–63 (1982).
219 Rehncrona, S.; Abdul-Rahman, A.; Siesjö, B.K.: Local cerebral blood flow in the postischemic period. Acta neurol. scand. *60:* suppl. 72, pp. 294–295 (1979).
220 Simon, R.P.; Griffiths, T.; Evans, M.C.; Meldrum, B.S.: Calcium overload in selectively vulnerable neurons of the hippocampus during and after ischemia: an electron microscopy study in the rat. J. cereb. Blood Flow Metab. *4:* 350–361 (1984).
221 Van Reempts, J.; Haseldonck, M.; Van de Ven, M.; Borgers, M.: Morphology and ultrastructural calcium distribution in the rat hippocampus after severe transient ischemia; in Bes, Graquet, Paoletti, Siesjö, Cerebral ischemia (Elsevier, Amsterdam 1984).
222 Hossmann, K.A.; Lechtape-Gruter, H.; Hossmann, V.: The role of cerebral blood flow for the recovery of the brain after prolonged ischemia. Z. Neurol. *204:* 281–299 (1973).
223 Snyder, J.V.; Nemoto, E.M.; Carroll, R.G.; Safar, P.: Global ischemia in dogs: intracranial pressures, brain blood flow and metabolism. Stroke *6:* 21–27 (1975).
224 Heffez, D.S.; Passonneau, J.V.: Effect of nimodipine on cerebral metabolism during ischemia and recirculation in the Mongolian gerbil. J. cereb. Blood Flow Metab. *5:* 523–528 (1985).
225 Steen, P.A.; Newberg, L.A.; Milde, J.H.; Michenfelder, J.D.: Nimodipine improves cerebral blood flow and neurologic recovery after complete cerebral ischemia in the dog. J. cereb. Blood Flow Metab. *3:* 38–43 (1983).
226 Steen, P.A.; Newberg, L.A.; Milde, J.H.; Michenfelder, J.D.: Cerebral blood flow and neurologic outcome when nimodipine is given after complete cerebral ischemia in the dog. J. cereb. Blood Flow Metab. *4:* 82–87 (1984).
227 Kazda, S.; Hoffmeister, F.; Garthoff, B.; Towart, R.: Prevention of the postischemia impaired reperfusion of the brain by nimodipine (Bay e 9736). Acta neurol. scand. *60:* suppl. 72, 302–303 (1979).
228 Steen, P.A.; Middle, J.H.; Michenfelder, J.C.: Cerebral metabolic and vascular effects of barbiturate therapy following complete global ischemia. J. Neurochem. *31:* 1317–1324 (1978).
229 Snyder, J.U.; Nemoto, E.M.; Carroll, R.G.; Safar, P.: Global ischemia in dogs: intracranial pressures, brain blood flow and metabolism. Stroke *6:* 21–27 (1975).
230 Schanne, F.A.X.; Kane, A.B.; Young, E.E.; Farber, J.L.: Calcium dependence of toxic cell death: a final common pathway. Science *206:* 700–702 (1979).
231 White, B.C.; Gadzinski, D.S.; Hoehner, P.J.; Krone, C.; Hoehner, T.; White, J.D.; Trombley, J.H.: Effect of flunarizine on canine cerebral cortical blood flow and vascular resistance post cardiac arrest. Ann. Emerg. Med. *11:* 119–126 (1982).
232 Karasawa, A.; Kumada, Y.; Yamada, K.; Shuto, K.; Nakamizo, N.: Protective effect of flunarizine against cerebral hypoxia-anoxia in mice an rats. J. Pharmacobiodyn. *5:* 295–300 (1982).
233 Newberg, L.A.; Steen, P.A.; Milde, J.H.; Michenfelder, J.D.: Failure of flunarizine to improve cerebral blood flow or neurologic recovery in a canine model of complete cerebral ischemia. Stroke *15:* 666–671 (1984).
234 Hossmann, K.A.; Paschen, W.; Csiby, L.: Relationship between calcium accumulation and recovery of cat brain after prolonged cerebral ischemia. J. cereb. Blood Flow Metab. *3:* 346–353 (1983).

235 Van Reempts, J.; Borgers, M.; Van Deal, L.; Van Eyndhoven, J.; de Ven, M.: Protection eith flunarizine against hypoxic-ischemic damage of the rat cerebral cortex. A quantitative morphologic approach. Archs Int. Pharmacodyn. Thér. *262:* 76–88 (1983).
236 Wauquier, A.; Ashton, D.; Clincke, G.; Van Reempts, J.: Pharmacological protection against brain hypoxia: the efficacy of flunarizine, a calcium entry blocker; in Amery, Clifford-Rose, Cerebral Hypoxia in the pathogenesis of migraine attacks, pp. 139–154 (Pitman, London 1982).
237 Turlapaty, P.D.; Altura, B.M.: Extracellular magnesium ions control calcium exchange and content of vascular smooth muscle. Eur. J. Pharmacol. *52:* 421 (1978).
238 Silverstein, F.S.; Buchanan, K.; Hudson, C.; Johnston, M.V.: Flunarizine limits hypoxia-ischemia induced morphologic injury in immature rat brain. Stroke *17:* 477–482 (1986).
239 Johnston, M.V.: Neurotransmitter alterations in a model of perinatal hypoxia-ischemic brain injury. Ann. Neurol. *13:* 511–518 (1983).
240 White, B.C.; Gadyinksi, D.S.; Hoehner, P.J.: Correction of canine cerebral cortical blood flow and vascular resistance after cardiac arrest using flunarizine, a calcium antagonist. Ann. Emerg. Med. *11:* 118–127 (1982).
241 Philis, J.W.; Delong, R.E.; Towner, J.K.: The effects of lidoflazine and flunarizine on cerebral reactive hyperemia. Eur. J. Pharmacol. *112:* 323–329 (1985).
242 Murphy, K.M.M.; Snyder, S.H.: Calcium antagonist receptor binding sites labelled with ^{3}H-nitrendipine. Eur. J. Pharmacol. *77:* 201–202 (1982).
243 Nachsen, D.A.; Blaustein, M.P.: The effects of some organic calcium antagonists on calcium influx in spre-synaptic nerve terminals. Molec. Pharmacol. *16:* 579–586 (1979).
244 Middlemiss, D.N.; Spedding, M.: A functional correlate for the dihydropyridine binding site in rat brain. Nature *314:* 94–96 (1985).
245 Turner, T.J.; Goldin, S.M.: Calcium channels in rat brain synaptosomes: identification and pharmacological characterization. J. Neurosci. *5:* 841–849 (1985).
246 Hofferberth, B.: Calcium entry blockers in the treatment of vertebrobasilar insufficiency. Eur. Neurol. *25:* suppl. 1, pp. 80–85 (1986).
247 Hofferberth, B.; Grotemeyer, K.H.: Flunarizine und Sulpirid – ein Vergleich zweier gebräuchlicher Antivertiginosa bei vertebrobasilärbedingtem Schwindel. Nervenarzt *56:* 553–559 (1985).
248 Scherer, H.; Schmidtmayer, E.; Hirche, H.: Die Wirkung von Bencyclan, Flunaricin und Naftidrofuryl auf den Nystagmus eines kalorischen Dauerreizes. Laryngol. Rhinol. Otol. *57:* 773–778 (1978).
249 Lee, J.A.; Watson, L.A.; Boothby, G.: Calcium antagonists in the prevention of motion sickness. Aviat. Space Environ Med. *75:* 45–48 (1986).
250 Hofferberth, B.: Die Anwendung von Flunarizin bei Patienten mit vertebrobasilärer Insuffizienz. Erfahrungen aus einer Pilotstudie. Arzneimittel-Forsch. *30:* 1817–1819 (1980).
251 Ell, J.; Gresty, M.: The effects of the 'vestibular sedative' drug, flunarizine upon the vestibular and oculomotor systems. J. Neurol. Neurosurg. Psychiat. *46:* 716–724 (1983).
252 Mangabeira-Albernaz, P.L.; Gananca, M.M.; Novo, N.F.; de Paiva, E.R.: Flunarizine and cinnarzine as vestibular depressants. A statistical study. ORL *40:* 92–100 (1978).

253 Oosterveld, W.J.: Flunarizine in vertigo. A double-blind placebo-controlled cross-over evaluation of a constant-dose schedule. ORL *44:* 72–80 (1982).

254 Theunissen, E.J.J.M.; Huygen, P.L.M.; Nicolasen, M.G.M.: Flunarizine treatment in dizzy patients with vestibular hyperreactivity and hyperventilation. ORL *48:* 203–208 (1986).

255 Theopold, H.M.: Nimodipin (Bay e 9736) ein neues Therapiekonzept bei Innenohrerkrankungen? Laryngol. Rhinol. Otol. *64:* 609–613 (1985).

256 Elbaz, P.; Haguenauer, J.P.; Muler, H.; Portmann, M.: Activité thérapeutique d'un inhibiteur de la surcharge calcique, la flunarizine, dans les troubles vestibulaires: étude à double-insu contre placebo. ORL *35:* 301–308 (1986).

257 Oosterveld, W.J.: Vestibular pharmacology of flunarizine compared to that of cinnarizine. ORL *36:* 157–164 (1974).

258 Schwerdtfeger, J.F.: Disturbed vestibular function. Ärztl. Prax. *30:* 451–452 (1978).

259 Turner, J.S.: The dizzy patient: diagnosis and treatments. Drugs *13:* 382–387 (1977).

260 Goldensohne, E.S.; Purpura, D.P.: Intracellular potentials of cortical neurons during focal epileptogenic discharges. Science *139:* 840–842 (1963).

261 Matsumoto, H.; Ajmone-Marsan, C.: Cortical cellular phenomena in experimental epilepsy: interictal manifestations. Expl. Neurol. *9:* 286–304 (1964).

262 Matsumoto, H.; Ajmone-Marsan, C.: Cortical cellular phenomena in experimental epilepsy: ictal manifestation. Expl. Neurol. *9:* 305–326 (1964).

263 Elger, C.E.; Speckmann, E.J.: Vertical inhibition in motor cortical epileptic foci and its consequences for descending activity to the spinal cord; in Speckmann, Elger, Epilepsy and motor system, pp. 152–160 (Urban & Schwarzenberg, München 1983).

264 Elger, C.E.; Speckmann, E.J.; Prohaska, O.; Caspers, H.: Pattern of intracortical potential distribution during focal interictal epileptiform discharges (FIED) and its relation to spinal field potential in the rat. Electroenceph. clin. Neurophysiol. *51:* 393–402 (1981).

265 Speckmann, E.J.; Caspers, H.: Paroxysmal depolarization and changes in action potentials induced by pentylenetetrazol in isolated neurons of *Helix pomatia.* Epilepsia *14:* 397–408 (1973a).

266 Speckmann, E.J.; Caspers, H.: Neurophysiologische Grundlagen der Provokationsmethoden in der Elektroenzephalographie. Z. EEG-EMG *4:* 157–167 (1973b).

267 Speckmann, E.J.; Caspers, H.; Janzen, R.W.: Relations between cortical DC shifts and membrane potential changes of cortical neurons associated with seizure activity; in Petsche, Brazier, Synchronization of EEG activity in epilepsies, pp. 93–111 (Springer, Wien 1972).

268 Li, C.L.; Van Buren, J.M.: Microelectrode recordings in the brain of man with particular reference to epilepsy and dyskinesia; in Somjen, Neurophysiology studied in man, pp. 49–63 (Excerpta Medica, Amsterdam 1972).

269 Prince, D.A.: Epileptogenesis in hippocampal and neocortical neurons; in Klee, Lux, Speckmann, Physiology and pharmacology of epileptogenic phenomena, pp. 151–162 (Raven Press, New York 1982).

270 Ayala, G.F.; Matsumoto, H.; Gumnit, R.J.: Excitability changes and inhibitory mechanisms in neocortical neurons during seizures. J. Neurophysiol. *33:* 73–85 (1970).

271 Speckmann, E.J.; Caspers, H.: Effects of pentylenetetrazol on isolated snail and mammalian neurons; in Chalazonitis, Boisson, Abnormal neuronal discharges, pp. 165–175 (Raven Press, New York 1978a).

272 Calvin, W.H.: Synaptic potential summation and repetitive firing mechanisms: input-output theory for the recruitment of neurons into epileptic bursting firing pattern. Brain Res. *39:* 71–94 (1972).

273 Klee, M.R.: Increased excitability of interneurons enhances positive feedback: possible mechanisms of epileptic activity; in Frigyesi, Subcortical mechanisms and sensorimotor activities, pp. 89–120 (Huber, Bern 1975).

274 Lux, H.D.: The kinetics of extracellular potassium: relation to epileptogenesis. Epilepsia *15:* 375–393 (1974).

275 Lux, H.D.; Heinemann, U.: Consequences of calcium-electrogenesis for the generation of paroxysmal depolarization shift; in Speckmann, Elger, Epilepsy and motor system, pp. 100–119 (Urban & Schwarzenberg, München 1983).

276 Heinemann, U.; Lux, H.D.; Marciani, M.G.; Hotmeier, G.: Slow potentials in relation to changes in extracellular potassium activity in the cortex of cats; in Speckmann, Caspers, Origin of cerebral field potentials, pp. 33–48 (Thieme, Stuttgart 1979).

277 Prince, D.A.; Wilder, B.J.: Control mechanisms in cortical epileptogenic foci. Archs Neurol. *16:* 194–202 (1967).

278 Gumnit, R.J.; Matsumoto, H.; Vasconetto, C.: DC activity in the depth of an experimental epileptic focus. Electroenceph. clin. Neurophysiol. *28:* 333–339 (1970).

279 Speckmann, E.J.; Caspers, H.; Janzen, R.W.C.: Laminar distribution of cortical field potentials in relation to neuronal activities during seizure discharges; in Brazier, Petsche, Architectonies of cerebral cortex, pp. 191–209 (Raven Press, New York 1978b).

280 Caspers, H.; Speckmann, E.J.: Cerebral pO_2 and pH: changes during convulsive activity and their significance for spontaneous arrest of seizures. Epilepsia *13:* 699–725 (1972).

281 Joebsis, F.F.; O'Connor, M.; Vitale, A.; Vreman, H.: Intracellular redox changes in functioning cerebral cortex. I. Metabolic effects of epileptiform activity. J. Neurophysiol. *34:* 735–749 (1971).

282 Kreisman, N.R.; Lamanna, J.C.; Rosenthal, M.; Sick, T.J.: Oxidative metabolic responses with recurrent seizures in rat cerebral cortex: role of systemic factors. Brain Res. *218:* 175–188 (1981).

283 Pumain, R.; Kurcewicz, I.; Louvel, J.: Fast extracellular calcium transients: involvement in epileptic processes. Science *222:* 177–179 (1983).

284 Binnie, C.D.: Flunarizine in epilepsy; in Vanhoutte, Paoletti, Gavoni, Calcium antagonists. Pharmacology and clinical research. Ann. N.Y. Acad. Sci. *522:* 710–711 (1988).

285 Blaustein, M.P.; Barbiturates block calcium uptake by stimulated and potassium depolarized rat sympathetic ganglia. J. Pharmac. exp. Ther. *196:* 80–86 (1976).

286 Pincus, J.H.; Lee, S.H.: Diphenylhydantoin and calcium: relation to epinephrine release from brain slices. Archs Neurol. *29:* 239–244 (1973).

287 Sohn, R.S.; Ferrendelli, J.A.: Anticonvulsant drug mechanism. Phenytoin, phenobarbital and ethosuximide and calcium flux in isolated presynaptic endings. Archs Neurol. *33:* 626–629 (1976).

288 Desmedt, L.K.C.; Niemeggers, C.J.E.; Janssen, P.A.J.: Anticonvulsant properties of cinnarizine and flunarizine in rats and mice. Arzneimittel-Forsch. *25:* 1408–1413 (1975).

289 Ashton, D.; Wauquier, A.: Effects of some anti-epileptic, neuroleptic and gabaminergic drugs on convulsions induced by *D,L*-allylglycine. Pharmacol. Biochem. Behav. *11:* 221–226 (1979).

290 Desmedt, L. K. C.; Niemeggers, C. J. E.; Janssen, P. A. J.: Anticonvulsant properties of cinnarizine and flunarizine in rats and mice. Arzneimittel-Forsch. *25:* 1408–1413 (1975).
291 Hoffmeister, F.; Benz, U.; Heise, A.; et al.: Behavioral effects of nimodipine in animals. Arzneimittel-Forsch. *32:* 347–360 (1982).
292 Meyer, F. B.; Anderson, R. E.; Sundt, T. M., Jr.; Sharbrough, F. W.: Selective central nervous system calcium channel blockers – a new class of anticonvulsant agents. Mayo Clin. Proc. *61:* 239–247 (1986).
293 Overweg, J.; Ashton, D.; de Beukelaar, F.; Binnie, C. D.; Wauquier, A.; van Wieringen, A.: Add-on therapy in epilepsy with calcium entry blockers. Eur. Neurol. *25:* suppl. 1, pp. 93–101 (1986).

Prof. Dr. A. Hartmann, Universitäts-Nervenklinik, Neurologie,
Sigmund-Freud-Strasse 25, D-5300 Bonn 1 (FRG)

Subject Index